Financial Literacy Education

Addressing Student, Business, and Government Needs

Financial Literacy Education

Addressing Student, Business, and Government Needs

EDITED BY **JAY LIEBOWITZ**

Harrisburg University of Science and Technology, Pennsylvania, USA

Foreword by Nan J. Morrison

CRC Press is an imprint of the
Taylor & Francis Group, an **informa** business
AN AUERBACH BOOK

CRC Press
Taylor & Francis Group
6000 Broken Sound Parkway NW, Suite 300
Boca Raton, FL 33487-2742

First issued in paperback 2018

ISBN-13: 978-1-4987-3853-8 (hbk)
ISBN-13: 978-1-138-89388-7 (pbk)

Library of Congress Cataloging-in-Publication Data

Financial literacy education : addressing student, business, and government needs / editor, Jay Liebowitz.
pages cm
Includes bibliographical references and index.
ISBN 978-1-4987-3853-8 (acid-free paper)
1. Financial literacy--Study and teaching. 2. Finance, Personal--Study and teaching. I. Liebowitz, Jay, 1957- editor.

HG179.F4628 2016
332.024--dc23 2015021701

Visit the Taylor & Francis Web site at
http://www.taylorandfrancis.com

and the CRC Press Web site at
http://www.crcpress.com

To anyone aspiring to be financially secure throughout
one's work life, retirement, and beyond

Contents

Foreword

When I stepped into the role of chief executive officer for the Council for Economic Education (CEE), I assumed the stewardship of an institution with an incredibly rich trove of experience, knowledge, and capacity in its field—bringing economic and personal finance education to grades K–12. And, yet, more than 65 years after our founding, achieving this mission across the national educational landscape is still a struggle. I would venture to describe it as a generational one. Educators and advocacy groups make some progress only to watch our country fall back into complacency, until another economic crisis rekindles interest in bringing this education to our children. The long view is a frustrating one.

Nonetheless, I think that we may have arrived at an "inflexion" point. Public awareness is certainly a plus, but it ebbs and flows; however, at this moment, parents resoundingly believe their children should learn these subjects. The involvement of business leaders and financial institutions, if it is constant and consistent, is an enormous aid, both in reaching constituents and in funding these efforts. Public/private partnerships have risen in esteem and usefulness, and there are simply more of us in the field to undertake them. There is desire and will, or political capital; and there is now enormous intellectual capital. It is my hope that undertakings such as this collection of chapters will help all of us identify efficient, successful, and stable paths forward, and I would like to share a little of our own experience.

While CEE's founders were concerned that America's youth would be able to successfully negotiate an increasingly competitive and complicated world market, they wrote primarily about ensuring the success of our democracy. In 1948, Ernest O. Melby, the dean of New York University School of Economics wrote: "Democracy will live if it works. Regardless of what democracy may do in the cultural and human relations area, if it fails on the economic front, it will most certainly go down in defeat. There is no kind of education more important than that which seeks to make the average American intelligent about our economic system and effective as a citizen in relation to it." In large measure, national strength and competitiveness still depend on the smallest unit, the minds and hearts of our children.

In my view, there are three key tools today for advancing and stabilizing this mission: first, educational technology gives us not only much greater reach in terms of bringing resources to more teachers, but is an important tool for keeping the national conversation alive; second, the adoption of standards and the passage of requirements are essential drivers because the overwhelming majority of teachers tell us that they only include these subjects when they are required*; and, finally, there is the painstaking work of bringing programs and resources to districts, tailored in ways that are useful, palatable, and sensible. CEE's experience on each of these fronts may be useful to discuss here.

Time is among the greatest adversaries for our teachers, both in terms of their own training and in terms of fitting more content into their days. They overwhelmingly endorse our ed/tech platform, EconEdLink, which provides them with lesson plans, access to other successful teachers, and ways to incorporate these lessons into other subjects. Teachers still value highly "face time" in the educational process. They want that hands-on experience and training; however, they prize the ease, access, and efficiency that educational technology can bring to themselves, as well as the engagement and enrichment that it can bring to their classrooms. CEE has been at this for some time, but we are now on the cusp of fully integrating our hands-on training with the considerable resources already available on our EconEdLink

* Eighty percent of teachers overall say that they teach economics because of a state requirement; 60% teach personal finance because of a state requirement.

site. Technology is expensive; so is hands-on training and the development of hard resources (textbooks). But technology is also scalable and therefore can reach more teachers and children.

On the subject of standards and requirements, CEE publishes its Survey of the States on a biennial basis. Published for the past 15 years, the Survey is an important benchmark for national progress. One might think the Survey is a simple collating task, and you could not be more wrong. In order to understand if a requirement is really a requirement, or just a vague suggestion, my staff reads each state's legislation to evaluate whether requirements are actionable. Currently, 43 states include personal finance in their K–12 standards, and 4 of these have even adopted the more robust national standards developed in 2014 under CEE's supervision. That is good news. But only 17 states require a course that includes personal finance for graduation, and only 6 states have any level of statewide testing. This is where the rubber meets the road.

Even where there is testing, it may be as little as one or two questions on an economics or civics examination, and, in many cases, what is tested is what is taught. Fortunately, there is now a nationally normed personal finance examination available on CEE's online assessment center. Whatever program a community decides to implement, this program's agnostic, standards-based examination allows them to assess students against a common bar.

Progress with regard to standards and requirements is all about state leadership and access to that leadership. Advocates for this education need feet on the ground; they need a presence in the school districts and the state houses. Fortunately, CEE's state affiliate structure has given us broad access across the country; but there is ample room on this playing field for more players.

This also brings me to the third key element in our experience—our statewide network of affiliates, which enables us to take our training and teaching resources and tailor them to the districts that we reach. In other words, this work still seems to advance best when it is adaptable on a district-by-district basis, "place-based" education. In Tennessee, for example, our programs unfolded as a result of engagement with their State Treasurer, who had decided that it would be better for personal finance education to start earlier. Our local affiliate worked with Treasurer Lillard to develop Smart Tennessee, a financial literacy program for K–8. It utilizes CEE's Financial Fitness for Life materials and provides these, as well as training and assessments for K–8 teachers. Students were tested, and the results were stellar—a 35% improvement in pre- to post-test scores. Importantly, the Tennessee Financial Literacy Commission is now reaching out to additional universities and school districts to expand the program, and it looks like other states are positioning themselves to emulate its success. Public/private partnerships, and partnerships between or among nonprofits are both essential and fruitful here.

A final element beyond the scope of most nonprofits is proof of mission effectiveness. Of course, we can test for student and teacher improvements after using our recommended resources. What is needed to keep the national conversation vibrant, however, are the kinds of evidence that the Financial Industry Regulatory Authority recently produced in an important new study: Young adults in three states with a financial education requirement had higher credit scores and fewer credit delinquencies than students in nearby states without those requirements.

Over my several years heading up CEE, I found that one of the biases against economic and personal finance education is that we are just about outputs, not outcomes—that we teach children to balance checkbooks, and not to build worthy lives. Of course, I do not want to diminish the value of a balanced checkbook. But what we are actually teaching children is how to determine, and how to understand, what "balance" means in their lives. Downturns and adjustments are necessary phases in economic cycles, but knowledge, prudence, and planning improve our chances for surviving hard times and prospering in good ones.

Our nation's young people should not be intimidated by finance or economics; they should view these subjects as enabling resources, which—along with character, family, and community—can help them harness opportunity on the horizon lines of their choice. Learning the grammar of economics gives young people the tools to recognize choice, the knowledge to evaluate options rationally, and the skill to act to fulfill their best dreams.

Nan J. Morrison
President and CEO
Council for Economic Education
New York, NY

Preface

"Financial literacy" is a term that has been used in many ways. According to the U.S. Financial Literacy and Education Commission (FLEC), financial literacy is the ability to use knowledge and skills to manage financial resources effectively for a lifetime of financial well-being (mymoney.gov). Over the years, federal/state governments, businesses, universities, and not-for-profit organizations have tried to better educate our children, young and mature adults, and seniors on financial literacy well-being. According to the 2011 U.S. National Strategy for Financial Literacy, four main goals were established:

1. Increase awareness of and access to effective financial education
2. Determine and integrate core financial competencies
3. Improve financial education infrastructure
4. Identify, enhance, and share effective practices

At the state-wide level, unfortunately, fewer than 20 U.S. states have required a course in some type of financial literacy for high school graduation. Many reports by such organizations as HigherOne, EdiFi, and universities have shown the statistical correlation between getting an early foundation in financial literacy and the likelihood of becoming a more financially responsible young adult (e.g., in terms of paying college loans on time). A key reference on the research into financial literacy is the Spring 2015 special issue on "Creating Financial

Capability in the Next Generation" in the *Journal of Consumer Affairs* (Blackwell).

Financial literacy should really be one of the key competencies that an individual learns throughout one's formal education. In keeping with this theme, this book is targeted at educators, practitioners, students, business leaders, and government officials to not only address some of the current work being done in financial literacy education but also stress some of the challenges and opportunities to further improve this area. We are fortunate to have some of the leading individuals and organizations showcase some of their work in this book. We are especially delighted to have Nan J. Morrison, president and CEO of the Council for Economic Education, write the Foreword in this book.

In my own experience, I was very surprised about seeing that college sophomores and juniors whom I taught weren't aware of such basic concepts as "compound interest" or how to calculate it. And certainly more advanced topics like capital budgeting techniques and retirement options were relatively foreign concepts to these students. Each student created 1-, 3-, and 5-year personal financial plans, and it helped to get them thinking about budgeting, saving, spending, and investing. Even identifying their "net worth" was an eye-opening experience.

Hopefully, this book will provide further clarity on what can be done to improve financial literacy awareness and education. Various approaches can be used from interactive games and tutorials to peer-to-peer mentoring to financial literacy contests to more formal education in these areas. This book will give you a sample of approaches and experiences in the financial literacy arena.

I thank all the contributors for their valuable work not only exhibited in this book but also in their quests to better educate students and adults in financial literacy nationwide and beyond. I also thank my colleagues and students at Harrisburg University of Science and Technology, as well as our "financial literacy" benefactor, Alex DiSanto, for supporting my work in financial literacy education. My deepest gratitude goes to my colleagues at Taylor & Francis for making this book a reality. And certainly, I am indebted to my family for their constant support and love, and for allowing me to implement and test-trial some of these financial literacy concepts over the many years.

Enjoy!

Jay Liebowitz, D.Sc.
Harrisburg University Distinguished Chair of Applied Business and Finance
(Formerly the DiSanto Visiting Chair in Applied Business and Finance)
Harrisburg, Pennsylvania

Contributors

Bryan Ashton
Assistant Director
Student Wellness Center
Office of Student Life
The Ohio State University
Columbus, Ohio

Cathy Faulcon Bowen
Professor of Agricultural and Extension Education
Consumer Issues
The Pennsylvania State University
University Park, Pennsylvania

Alicia Puente Cackley
Director
Financial Markets and Community Investment
Government Accountability Office
Arlington, Virginia

Patricia W. Collins
Director
Student Money Management Center
Sam Houston State University
Huntsville, Texas

Theodore R. Daniels
President
Society for Financial Education and Professional Development
Washington, DC

Eric R. Heckman
President
Financial Knowledge Institute
San Jose, California

Hilary Hunt
President
Hilary Hunt Education Consulting
Harrisburg, Pennsylvania

Rakhee Jain
Co-President
Moneythink—University of Chicago Chapter
Chicago, Illinois

Mary Johnson
Vice President
Financial Literacy and Student Aid Policy
Higher One, Inc.
New Haven, Connecticut

Jay Liebowitz
Harrisburg University Distinguished Chair of Applied Business and Finance
Harrisburg University of Science and Technology
Harrisburg, Pennsylvania

Amy Marty
Program Manager
National Endowment for Financial Education
Denver, Colorado

Bonnie T. Meszaros
Associate Director
Center for Economic Education and Entrepreneurship
University of Delaware
Newark, Delaware

John Pelletier
Director
Center for Financial Literacy
Champlain College
Burlington, Vermont

Mary C. Suiter
Assistant Vice-President
Research-Economic Education
Federal Reserve Bank of St. Louis
St. Louis, Missouri

PART I

K–12 Focused Financial Literacy Education

1

The Changing Landscape of K–12 Personal Finance Education

BONNIE T. MESZAROS AND MARY C. SUITER*

Contents

Introduction

For more than a decade, there has been an increased interest in the inclusion of personal finance in the K–12 curriculum. A number of factors have fueled the discussion including statistics on the impact of poor financial decisions made by young adults, the high levels of consumer debt, in particular college loan debt, predatory lending, and expanded access to credit for younger populations.

Articles from the popular press and the academic community have provided more perspective and insight. The two previous chairmen

* The views expressed in this article are those of the authors and not those of the Federal Reserve Bank of St. Louis or the Federal Reserve System.

of the Federal Reserve, Alan Greenspan (2001) and Ben Bernanke (2006a,b), and the creation of the President's Advisory Council on Financial Literacy and the Financial Literacy and Education Commission continue to bring attention to the importance of a financially literate citizenry. Potential implementation of these efforts has been reinforced by the Consumer Financial Protection Bureau's (CFPB, 2015a) release of a resource guide aimed to aid policymakers to further the development and implementation of financial education. CFPB Director Richard Cordray stated: "Financial education in our schools is critical to the financial well-being of future generations" (CFPB, 2015b).

Most agree that achievement of this goal rests in part with the K–12 school systems. However, some feel that financial literacy starts at home. Financial expert Bill Hardekopf, chief executive officer of lowcards.com, said that school-based financial education is great but that financial education is the parents' responsibility (Engel, 2015). Unfortunately, not all parents provide their children with sound financial habits. The Charles Schwab's 2011 Teen and Money Survey found that parent discussions are not necessarily translated into knowledge about financial tools. When surveyed about topics their parents discuss with them, smart money management was near the bottom, along with conversations about drugs, alcohol, dating, and sex.

High school graduates should be grounded in the basics of personal finance and possess the skills and knowledge necessary to make informed decisions. Today's graduates will be expected to take more responsibility for their financial well-being. They will be faced with an array of complex financial services and sophisticated products. Without knowledge and skills in personal finance, making rationale, informed decisions on the use of their money and planning for future events, such as retirement, will be difficult.

The second National Financial Capability Study (2013) surveyed adults on a variety of topics. When asked whether they thought financial education should be taught in the schools, 89% of the respondents said yes—a key indication that Americans recognize the importance of financial capability and the role schools can play in preparing their students. One of the conclusions from the study was that ensuring citizens have access from an early age to adequate information should be a high priority for policy makers and for society as a whole.

Status of Financial Education in Schools

Despite this push for financial education in the schools, the reality is not promising. The 2014 Survey of the States conducted by the Council for Economic Education found 43 states, three fewer than in the 2011 survey, included personal finance in their standards, but only 35 states required that these standards be implemented (Council for Economic Education, 2014). Even fewer states, 19, required that a high school course be offered, and only 17 required that a course be taken. The most discouraging results were that only six states required student testing of personal finance concepts. Having standards and offering courses does not necessarily lead to change in student knowledge and capabilities and does not move us closer to achieving the goal of financial education for all students. To ensure that personal finance courses are taught well and include quality content requires testing.

The 2011 Center for Financial Literacy at Champlain College (see Chapter 2) used national data* to grade states on their efforts to produce financially literate high school graduates. Sixty percent of states received grades of C or less, and of these 44% had grades of D or F. States receiving an A require a stand-alone personal finance course or that personal finance topics be taught as part of another mandatory course and require that student knowledge is assessed. States receiving an F have few requirements, or none at all, for personal finance education.

The results of the first 2012 PISA (Programme for International Student Assessment) Financial Literacy Assessment also show that more needs to be done. This assessment was designed to measure the proficiency of 15-year-old students in demonstrating and applying the personal finance knowledge and skills learned both in and out

* Many state standards were examined *as* well as the Jump$tart Coalition for Personal Finance Literacy's *National Standards in K–12 Personal Finance Education*, the Department of the Treasury's "Financial Education Core Competencies," and PISA's 2012 "Financial Literacy Assessment Framework."

of school. The definition of financial literacy used in the PISA 2012 Financial Literacy Assessment Framework is:

> Financial literacy is knowledge and understanding of financial concepts and risks, and the skills, motivation and confidence to apply such knowledge and understanding in order to make effective decisions across a range of financial contexts, to improve the financial well-being of individuals and society, and to enable participation in economic life. (p. 13)

Eighteen countries and economies including the United States participated in the assessment. The United States ranked ninth out of 18 countries—behind Latvia, Poland, Czech Republic, New Zealand, Australia, Estonia, Flemish Community (Belgium), and Shanghai (China).

What Is Being Taught and Where

Is Personal Finance a Homeless Curriculum?

In *Social Education* (2005), John Morton described personal finance as a "homeless curriculum," because, at the time, there was no set of standards that defined the discipline and informed the personal finance curriculum and because personal finance has no true academic "home." Because there is no set of standards or principles that defines the discipline and informs the personal finance curriculum, personal finance has no true academic "home." It is taught in whatever high school department the school or school district chooses. It could be taught in the social studies department in one school, in the business department at a school a few miles away, in the family and consumer science department across the state, and perhaps not taught at all in a school across the state line.

When personal finance is taught, states or local school districts for the most part determine what to teach, at what grade levels, and in what courses. This scattered approach makes it difficult to assess the content and quality of what is being taught across the nation. Although the Financial Literacy and Education Commission developed a national strategy for promoting financial education, there is no national curriculum for states or school districts to follow. Another concern is that those teaching personal finance may never have had a course in personal finance. In Missouri, for example, any certified teacher—regardless of his or her subject area—is qualified to teach the high school personal finance course.

Science is not taught in the social studies department by teachers trained to teach history and geography or by mathematics teachers. And students do not take one science course in high school that covers everything they should know about science. And we do not omit testing and other assessments of student knowledge in science. But, more often than not, this is exactly what we do with financial education.

Integration across the Curriculum

Another concern is the integration of personal finance into other subjects. In elementary, middle, and (often) high schools, if personal finance is taught at all, it is integrated into a variety of courses. In a 2010 survey, two-thirds of the states that reported having personal finance standards indicated that personal finance was being integrated into courses in a variety of content and subject areas (Hill and Meszaros, 2011). Research in economic education has shown that the integration of economics into the teaching of other disciplines may be insufficient. Little knowledge is gained when economics is integrated into a social studies or consumer economics course (Walstad and Soper, 1989). Students need a capstone economics course to fully understand and grasp the content. Although it doesn't seem to be a big leap to assume the same is true for personal finance instruction, research is needed to see if the weaving of personal finance instruction across the curriculum is as effective as a stand-alone personal finance course.

Given this uneven and marginalizing approach to financial education, it is not surprising that some research suggests that it doesn't work in its current form. But is that a fair conclusion to draw? What if the same approach were taken with personal finance education that is taken with other disciplines? What if personal finance instruction began in the early grades and followed a scope and sequence that added depth and breadth of content and then culminated with a high school course?

More people are beginning to consider the importance of starting financial education early. The National Association of State School Boards of Education's Commission on Financial and Investor Literacy has recommended the integration of financial education beginning in kindergarten and progressing across the grades. The argument can be made that starting early is essential (Suiter and Meszaros, 2005). A Research Brief from the Center for Financial Security at the University of Wisconsin–Madison (2012) outlines the literature related to student learning of economic and financial concepts. This literature suggests that children make great strides in learning economic and financial concepts by the age of 12 that may be attributed to cognitive development related to their age—for example, developing the ability to understand multiple causation or arithmetic. However, other studies point out that children's direct experience and socialization—including teaching—are important, too.

We can demonstrate the need for early education based on current survey information. According to a 2012 report by Digitas, children between the ages of 6 and 12 have more than a trillion dollars in buying power (Robinson, 2012). This includes what they buy and what they influence their parents to spend on them. This reinforces the notion that children need the knowledge and skills to make informed decisions about managing their money at an early age. Children bring their experiences with money into the classrooms. Some of their learnings are correct and some are not. Therefore, teachers must be able to understand and correct the misconceptions so they do not persist as students advance through the grades (Soper and Brenneke, 1987; Schug, 1993–1994; John, 1999; Sherraden, Johnson, Guo, and Elliott, 2011).

National Standards and Assessment for Financial Literacy

An important step toward providing an academic home for personal finance and ensuring that it is taught throughout the K–12 sequence is the development of standards and assessments. Standards provide guidance regarding what to teach, when to teach it, and how to assess it. Until recently, there were no national standards framed in the economic way of thinking—that is, the application of a decision-making process or cost/benefit analysis across personal finance content. Nor were existing standards linked to a nationally normed test to provide guidance. The idea for a set of national personal finance standards that deepen student understanding and require the application of cost/benefit analysis emerged at a conference, "Assessment and Evaluation of K–12 Personal Finance and Economic Education in the 21st Century: Knowledge, Attitudes, and Behavior," held at the Federal Reserve Bank of St. Louis in May 2011. After the meeting, the Council for Economic Education formed a writing committee. After reviewing several* and gaining a perspective from a variety of experts, the committee agreed on six standards: earning income, buying goods and services, saving, using credit, financial investing, and protecting and insuring.

The final document, "National Standards for Financial Literacy" (CEE, 2013), represents the thinking of the writing committee informed by input from three additional committees: a group of economic and financial literacy educators made up of individuals who provide professional development for K–12 teachers; a review committee of nationally recognized experts from academic institutions, the Federal Reserve, and business; and an educator review committee consisting of teachers plus input from a variety of groups such as the National Association of Economic Educators and the National Association of State Boards of Education and feedback from a presentation sponsored by the American Economic Association's Committee on Economic Education.

The National Financial Literacy Standards (CEE, 2013) have several characteristics that make them unique:

- Have an economic perspective with a goal of producing materials that are practical but academically sound.
- Do not assume prior knowledge in economics or personal finance.
- Use nontechnical language to offer guidance for the inexperienced teacher with limited personal finance and economics training; do not use the language of the discipline.
- Begin with economic decision-making as opposed to the rules of thumb or cookbook recommendations often found in the mainstream media.

---

* Data sources included the financial literacy legislation summaries from the National Conference of State Legislatures, the online data on state financial education requirements from the Jump$tart Coalition on Personal Financial Literacy, the 2011 Survey of the States produced by the Council for Economic Education, and research on individual states when inconsistencies existed.

- Include benchmarks reflecting the thinking of behavioral economics.
- Include financial capabilities.
- Include benchmarks for grades 4, 8, and 12 to provide guidance on what content should be taught when and how it can be taught with increasing degrees of sophistication.
- Provide many examples of the financial decisions people face in their daily lives.

The writing committee envisioned that the standards would result in better curriculum development and alignment leading to better teaching of important content and to the development of high-quality standards-aligned assessments.

Once the National Standards for Financial Literacy were complete, the next step was to develop a set of nationally normed tests aligned to the standards. Again, the Council for Economic Education created a committee that developed three tests for high school, middle school, and upper elementary school. These tests were field tested in the spring of 2015 and are available on the Council's website. The tests along with the standards allow schools and districts to address K–12 personal finance, adding depth and breadth each year, and to assess student progress.

The personal finance tests along with the National Standards for Financial Literacy also provide an opportunity for research on program effectiveness that do not rely on tests that are linked to a specific curriculum package or program. One of the discouraging aspects of financial education has been the lack of this type of assessment.

Hathaway and Khatiwada (2008) found that most research fails to demonstrate the effectiveness of financial education programs. They stated two possible reasons for this. Programs are ineffective in transferring knowledge because of their design or how they are administered or the programs are not being evaluated properly. Their findings were based on adult programs but have implications for evaluation of K–12 programs. They concluded that a framework for all types of financial literacy programs was needed. A similar conclusion was reached by Fox, Bartholomae, and Lee (2005). They recommended using Jacobs's (1988) five-tiered approach to evaluation. Walstad, Rebeck, and MacDonald (2010) outlined the five steps for use in evaluating the video-based curriculum, *Financing Your Future* (Emery and Suiter, 2007). These steps for financial literacy included:

1. Setting a clear definition of content
2. Training teachers in both content and the curriculum or program materials
3. Using reliable and valid instruments to assess specified knowledge outcomes
4. Collecting pre- and post-test data
5. Analyzing data using appropriate forms of statistical analysis

The National Standards for Financial Literacy provide the framework for K–12 personal finance education that Hathaway and Khatiwada recommended. The development of the three personal finance tests aligned with these standards create the reliable and valid instruments required for step 3 and the multiple test versions for pre-/post-test purposes in step 4.

Teacher Education

Those who point to the failure of financial education in schools often neglect an important aspect of any school-based instruction—that is, whether teachers are prepared to teach the content. High school teachers receive education specific to their disciplines. And, throughout their careers, they take additional education in that discipline through in-service programs, graduate-level courses, seminars, and other programs. Elementary and middle school teachers have content classes (although not as many as those of high school teachers) while seeking degrees. Additionally, these teachers take a certification test—often the PRAXIS—that tests their content knowledge. They also have opportunities to attend in-service professional development to augment their education in specific disciplines.

The same cannot be said for those who teach personal finance. First, as discussed earlier, personal finance is a "homeless" curriculum. It might be taught in business, family and consumer science, social studies, or mathematics departments at the high school and middle school levels. Teachers in these disciplines are not prepared to teach the full range of topics covered in a personal finance course. Loibl (2008) found in Ohio, for example, that business teachers were more comfortable teaching about investment and not very comfortable teaching about budgeting. On the other hand, family and consumer science teachers were more comfortable with budgeting and not very comfortable with investments. And although math teachers have a great grasp of the mathematics associated with personal finance, they do not have adequate understanding of the content of the discipline. Social studies teachers receive no training in personal finance content while in school. Likewise, elementary and middle school teachers receive no instruction in personal finance. And the PRAXIS examination doesn't cover personal finance. You can't teach what you don't know, and current research regarding the success of personal finance education supports this fact.

Internationally, the Organisation of Economic Co-operation Development's Program for International Student Assessment (PISA) (2014) tested 15-year-olds on their knowledge of personal finance and ability to solve financial problems. This was the first large-scale international study undertaken to assess financial literacy among young people. Students from 18 countries participated. Four of the top nine countries in which students performed the best on the PISA assessment—Flemish Community (Belgium), Czech Republic, Latvia, and the United States—also had high percentages of students in schools where teachers received professional development in financial education.

In a report titled, "What Works," Mike Watts (2005) described factors that have been identified through research as contributing to student learning of economics. Among those factors, the most important are students' skills and teachers' training in economics or economic education. Good instructional materials also matter, as does the amount of time teachers spend teaching economics.

After student ability, teachers' ability—that is, their knowledge of the subject and their attitudes about the subject—is most positively related to student learning. In 2005, Baron-Donovan, Wiener, Gross, and Block-Lieb (2005) investigated the topic of teacher training as a component of successful delivery of financial education, based on a 2-day train-the-trainer program with multiple measures. The authors sought to demonstrate whether individuals from diverse backgrounds are prepared to teach basic financial literacy. They used a combination of focus groups and a pre- and post-testing survey to determine increases in teacher satisfaction, confidence, and motivation. The survey included 16 financial knowledge questions and 14 attitude measures. Participants had an average pre-test knowledge score of 81% and an average post-training knowledge score of 90%. In addition, participants showed the desired changes in attitude for nearly half of the attitude items. The authors concluded that teachers gained not only an understanding of what to teach but also the confidence to teach complex content, and knowledge of materials they can use to address the content in the classroom.

These results suggest that teacher training matters. Loibl (2008) also addressed teacher confidence for high school financial education programs in Ohio. Her study addressed (1) teacher confidence in the larger disciplines within which the topic of financial education is often addressed (i.e., math, social studies, economics, family and consumer science, and business education); (2) teacher strategies for gathering personal finance information; and (3) teacher knowledge about personal finance. Loibl's survey included a short quiz on financial knowledge. Teachers from almost all disciplines struggled. This suggests a need to provide training for personal finance educators.

Way and Holden (2010) conducted a 2-year study on the background and capacity of teachers to teach personal finance. They addressed whether teachers were competent and confident when it came to teaching personal finance. They found that

- Eighty percent of states have adopted personal finance education standards or guidelines of some kind. This is an increase from 42% reported in 1998.
- Overall, 63.8% of teachers did not feel qualified to use their state's financial literacy standards.
- Eighty-nine percent of teachers agreed or strongly agreed that students should take a financial literacy course or pass a test for high school graduation.
- Only 29.7% of teachers were teaching any financial education, whether in existing classes or special classes on finance topics.
- Only 37% of K–12 teachers took a college course in personal finance.
- Only 11.6% of K–12 teachers took a workshop on teaching personal finance.
- Less than 20% of teachers reported feeling very competent to teach personal finance.

Personal Finance Instructional Materials

Those charged with teaching personal finance at the K–12 level have a plethora of materials from which to choose. For example, the Jump$tart Coalition website lists more than 800 financial education resources, ranging from print and online materials to videos, DVDs, games, and other formats. The President's Advisory Council on Financial Capability recommended the development of a website, Money As You Learn, that offers personal finance lessons for use in mathematics and English classes. Some materials are free or inexpensive and are marketed by a specific organization or business. Some are high quality with accurate content and appropriate pedagogy. Others push a particular product, promote a particular philosophy, and have incorrect content.

The task of selecting quality, grade-appropriate materials linked to state and national standards can be overwhelming. The National Association of Economic Educators recognized that school districts, teachers, and those responsible for personal finance education need a tool to evaluate instructional lessons and curriculum packages, so they worked with representatives from the academic community and the Federal Reserve Banks to design a rubric for evaluating materials. The committee produced four rubrics for use with textbooks, curriculum packages, individual lessons, and nonlesson resources such as videos, activity books, and podcasts. These were field tested by teachers and economic educators and revised; they are available free of charge on the National Association of Economic Educators website. Another possible resource is mymoney.gov, a web site sponsored by the U.S. Financial Literacy and Education Commission.

Promising Practices and Results

Recent research on the impact of financial education and student knowledge and financial capabilities has been promising. Asarta, Hill, and Meszaros (2014) examined the impact of teacher education along with the implementation of a well-designed curriculum on student learning. They looked at secondary level education in Delaware, Pennsylvania, and New Jersey, where teachers attend Keys to Financial Success training. They found that training combined with quality curriculum significantly improved the average personal finance knowledge of students. Swinton, DeBerry, Scafidi, and Woodard (2007) found that high school students whose teachers attended personal finance workshops scored better on the statewide assessment than students whose teachers did not attend workshops. And the more workshops the teachers attended, the better the students scored. Harter and Harter (2012) found that teacher participation in personal finance workshops or in a graduate course in personal finance led to improved student scores on a multiple-choice test covering financial concepts. Students of teachers who participated in the graduate courses had improved slightly more than students of teachers who participated in a workshop. Walstad, Rebeck, and MacDonald (2010) also found that students who were taught by trained teachers using the *Financing Your Future* package gained statistically significantly more knowledge than their peers in a comparison group.

Fewer (and smaller) studies are available at the elementary grades. However, the results from these studies are positive: There is some evidence that starting earlier and building on student learning would improve outcomes. Sosin and her colleagues (1997) found similar gains on an economics test given across several elementary and middle school grades. They concluded that the older students could have achieved more cumulative learning if the curriculum had been introduced in earlier grades. A study by Berti and Monaci (1998) found that students who received 20 hours of instruction about banks scored higher in interviews than those who did not receive the instruction. And the students maintained this knowledge even 4 months after the curriculum was completed. A 2008 study by Smith, Sharp, and Campbell used pre- and post-tests and found a significant improvement among a group of 160 students who were taught using a financial literacy curriculum (Smith, Sharp, and Campbell, 2011). There was no control group for this study. A study by Grody, Grody, Kromann, and Sutliff (2008) involved a single third-grade classroom that was divided into control and treatment groups. The treatment group was read a children's book containing financial content. The treatment group showed gains from pre-test to post-test. Sherraden, Johnson, Guo, and Elliott (2011) found in a very small study that elementary students who participated in the I Can Save program (which included enrollment in a matched savings program along with instruction in financial education) scored significantly higher on a financial literacy test administered in the fourth grade than those who did not participate in the program.

Less research has been done on financial capabilities at the K–12 level. However, some promising findings on the impact of personal finance instruction in the schools have emerged. A 2001 study by Bernheim, Garret, and Maki found that state mandates for financial education not only significantly increase students' exposure to personal finance content, but also increase the rate at which individuals save and accumulate wealth as adults. The authors also point out that there are implementation lags related to the mandates that have to do with teachers and students adjusting to changes in the curriculum.

A more recent Financial Industry Regulatory Authority Foundation–funded study, "State of financial education mandates; It's all in the implementation," supports these results (Urban, Schmeiser, Collins, and Brown, 2015). The researchers found that a rigorous financial education program that is carefully implemented in classrooms can result in improved credit scores and lower probability of credit delinquency among young adults. The study compared the credit behaviors for young adults in Georgia, Idaho, and Texas—three states that implemented financial education requirements after 2000—to the credit behaviors of young adults in adjacent states that did not have personal finance education mandates. The study found that 3 years after the financial education mandate was implemented, young adults in those states had significant increases in credit scores and lower delinquency rates on credit accounts than young adults in the nonmandate states. The authors point out that it takes time for financial mandates to have an impact. Their results show an implementation lag: the positive effects 1 year after implementation were few; however, by the second year after implementation, the results were consistently positive.

The Arizona Pathways to Life Success for Undergraduate Students is a longitudinal study of young adults' changing financial knowledge and practices. The study involves surveying a cohort during their first year at the University of Arizona in 2007 and resurveying them through middle age. The researchers are studying the relationship between college financial behaviors and adult financial capability in order to understand how early financial behaviors contribute to success and well-being in later life. Each wave of the study results in more understanding. The authors of the first wave of the study (Soyeon and Serido, 2009) suggest that today's complex financial services industry and the emphasis placed on personal responsibility for financial security mean that young people need financial education now more than ever. The results of the study indicate that repeated exposure to financial concepts makes the most powerful contribution to financial capability. This supports an approach for financial education such as that used with traditional school subjects—a scope and sequence that begins in early grades and builds throughout students' school careers. The study goes on to suggest that continued exposure does more than increase financial knowledge about personal finance. Each exposure provides the potential to encourage lifelong learning in personal finance—encouraging students to seek out financial workshops, classes, books, and articles.

The second wave of the study (Soyeon and Serido, 2011) further supports the importance of ongoing financial education. Those participants with cumulative financial education knew more and reported more positive financial behaviors. In this wave of the study, researchers also documented a "snowball effect" of financial education: "earlier financial education exponentially increases the likelihood of later financial education." This includes continuing formal education but also informal education, such as books, magazines, and seminars.

Conclusion

Education is an investment in human capital. Its goal is to prepare students to lead productive and fulfilling lives and to develop as lifelong learners. This benefits not only the individual student, but society as a whole. As a result, K–12 education covers a wide array of important content and essential skills, such as reading, writing, history, literature, mathematics, and economics. All of this is included because students need the skills and the rich content background from these disciplines to develop and function as adults. The same can be said for including personal finance as well.

Lusardi and Mitchell (2014) suggest that financial knowledge is a specific type of human capital and that those who acquire it will be rewarded with above-average expected returns on their investments. Furthermore, Lusardi, Michaud, and Mitchell (2012), using a life-cycle model, show that financial knowledge is a key factor in wealth inequality. These results only strengthen the argument that all students need personal finance education. To achieve this, personal finance instruction must be mandated by state standards that are tested and offered throughout the K–12 curriculum by knowledgeable teachers using high-quality instructional materials.

References

Asarta, C. J., Hill, A. T., and Meszaros, B. T. (2014). The features and effectiveness of the keys to financial success curriculum. *International Review of Economics Education*, *16*, 39–50.

Baron-Donovan, C., Wiener, R., Gross, K., and Block-Lieb, S. (2005). Financial literacy teacher training: A multiple-measure evaluation. *Financial Counseling and Planning*, *16(2)*, 63–75.

Bernanke, B. S. (2006a). At the Conversation with the Chairman: A Teacher Town Hall Meeting, Federal Reserve Board of Governors, Washington, DC. Retrieved from http://www.federalreserve.gove/newsevents/speech/bernanke20120807a.htm.

Bernanke, B. S. (2006b). *Financial Literacy. Testimony before the Committee on Banking, Housing, and Urban Affairs of the United States Senate.* Retrieved from http://www.federalreserve.gov/newsevents/testimony/Bernanke20060523a.htm.

Bernheim, B. D., Garrett, D. M., and Maki, D. M. (2001). Education and saving: The long-term effects of high school financial curriculum mandates. *Journal of Public Economics*, *80(3)*, 435–465.

Berti, A. E., and Monaci, M. G. (1998). Third graders' acquisition of knowledge of banking: Restructuring or accretion? *British Journal of Educational Psychology*, *68(3)*, 357–371.

Center for Financial Security, University of Wisconsin–Madison. (2012). *Research briefing: Youth, financial literacy, and learning: The role of in-school financial education in building financial literacy.*

Charles Schwab Corporation. (2011). *Teen and money survey.* Retrieved from http://www.schwabmoneywise.com/public/moneywise/calculators_tools/families_money_surveys/teens_money_survey.

Consumer Financial Protection Bureau. (2015a). Press Release. *Advancing K–12 financial education: A guide for policymakers.* Retrieved from http://www.consumerfinance.gov/reports/advancing-k-12-financial-education-a-guide-for-policymakers/.

Consumer Financial Protection Bureau. (2015b). Press Release. *CFPB launches nationwide effort to advance financial education in schools.* Retrieved from http://www.consumerfinance.gov/newsroom/cfpb-launches-nationwide-effort-to-advance-financial-education-in-schools/.

Council for Economic Education. (2013). *National standards for financial literacy.* New York: Council for Economic Education.

Council for Economic Education. (2014). *Survey of the states: Economic and personal finance education in our nation's schools.* New York: Council for Economic Education. Retrieved from http://www.councilforeconed.org/policy-and-advocacy/survey-of-the-states/.

Emery, B., and Suiter, M. (2007). *Financing your future.* New York: National Council on Economic Education.

Engel, M. (2015). Students taking financial education courses have better credit. *New York Daily News.* Retrieved from http://www.nydailynews.com/life-style/students-financial-education-better-credit-stud-article-1.2106166?cid=bitly.

FINRA Investor Education Foundation. (2013). *National financial capability study.* Retrieved from http://www.usfinancialcapability.org/downloads/NFCS_2012_Report_Natl_Findings.pdf.

Fox, J., Bartholomae, S., and Lee, J. (2005). Building the case for financial education. *Journal of Consumer Affairs, 39,* 195–214.

Greenspan, A. (2001). Remarks by Chairman Alan Greenspan: The importance of education in today's economy. Speech presented at the Community Affairs Research Conference of the Federal Reserve System. Retrieved from http://www.federalreserve.gov/boarddocs/speeches/2001/20011026/default.htm.

Grody, A., Grody, A., Kromann, E., and Sutliff, J. (2008). *A financial literacy and financial services program for elementary school grades—Results of a pilot study.* Retrieved from http://ssrn.com/abstract=1132388.

Harter, C. L., and Harter, J. F. R. (2012). Does a graduate course in personal finance for teachers lead to higher student financial literacy than a teacher workshop? *Journal of Consumer Education, 29,* 35–46. Retrieved from http://www.cefe.illinois.edu/JCE/archives/vol 29.html.

Hathaway, I., and Khatiwada, S. (2008). Do financial education programs work? *Federal Reserve Bank of Cleveland Working Paper No. 08-0.* Retrieved from http://www.usc.edu/dept/chepa/IDApays/publications/do_financial_education.pdf.

Hill, A., and Meszaros, B. (2011). Status of K–12 personal financial education in the United States. *Journal of Consumer Education, 28,* 1–15. Retrieved from http://www.cefe.illinois.edu/JCE/archives/vol28.html.

Jacobs, F. (1988). The five-tier approach to evaluation: Context and implementation. In Weiss, H., & Jacobs, F. (Eds.). *Evaluating Family Programs* (36–68). New York: Aldine deGruyter.

John, D. R. (1999). Consumer socialization of children: A retrospective look at twenty-five years of research. *Journal of Consumer Research, 26(3),* 182–213.

Loibl, C. (2008). *Survey of financial education in Ohio's schools: Assessment of teachers, programs, and legislative efforts.* Ohio State University P-12 Project. Retrieved from https://education.ohio.gov/getattachment/Topics/Academic-Content-Standards/Financial-Literacy-and-Business/Financial-Literacy/Loibl_ExecutiveSummary_print-1.pdf.aspx.

Lusardi, A., and Mitchell, O. S. (2014). The economic importance of financial literacy: Theory and evidence. *Journal of Economic Literature, American Economic Association, 52(1)*, 5–44.

Lusardi, A., Michaud, P., and Mitchell, O. S. (2012). *Optimal financial knowledge and wealth inequality*. Global Financial Literacy Education Center. George Washington University. Retrieved from http://gflec.org/wp-content/uploads/2014/12/optimal-financial-knowledge-and-weathly-inequality.pdf.

Morton, J. S. (2005). The interdependence of economic and personal finance education. *Social Education, 69(2)*, 66–69.

OECD. (2014). *PISA 2012 Results: Students and money: Financial literacy skills for the 21st century* (Volume VI), PISA, OECD Publishing. Retrieved from http://dx.doi.org/10.1787/9789264208094-en.

Robinson, J. (2012). *The next generation of consumers*. Bostino. Retrieved from http://bostinno.streetwise.co/channels/the-next-generation-of-consumers/.

Schug, M. (1993–1994). How children learn economics. *The International Journal of Social Education, 8(3)*, 25–34.

Sherraden, M. S., Johnson, L., Guo, B., and Elliott, W. (2011). Financial capability in children: Effects of participation in a school-based financial education and savings program. *Journal of Family & Economic Issues, 32(3)*, 385–399.

Smith, R. C., Sharp, E. H., and Campbell, R. (2011). Evaluation of Financial Fitness for Life program and future outlook in the Mississippi Delta. *Journal of Economics and Economic Education Research*, 12, 25–39.

Soper, J., and Brenneke, J. (1987). Economics in the school curriculum. *Theory into Practice: Developing Economic Literacy, 26(3)*, 183–189.

Sosin, K., Dick, J., and Reiser, M. L. (1997). Determinants of achievement of economics concepts by elementary school students. *The Journal of Economic Education, 28(2)*, 100–121.

Soyeon, S., and Serido, J. (2009). *Arizona pathways to success for university students, cultivating positive financial attitudes and behaviors for healthy adulthood*. Retrieved from http://tcainstitute.org/UA%20APlus%20report%20v9.pdf.

Soyeon, S., and Serido, J. (2011). *Young adults financial capabilities wave 2 Arizona Pathways to success for university students*. Retrieved from http://tcainstitute.org/APLUS-Wave-2-Report.pdf.

Suiter, M., and Meszaros, B. (2005). Teaching about saving and investing in the elementary and middle school grades. *Social Education, 69(2)*, 92–95.

Swinton, J., DeBerry, T. W., Scafidi, B., and Woodard, H. C. (2007). The impact of financial education workshops for teachers on students' economics achievement. *The Journal of Consumer Education, 27*, 63–77.

Urban, C., Schmeiser, M., Collins, M. J., and Brown, A. (2015). State of financial education mandates; It's all in the implementation. *Insights: Financial Capability*. Washington, DC: FINRA Investor Education Found. Retrieved from https://www.finra.org/sites/default/files/investoreducation foundation.pdf.

Walstad, W., and Soper, J. (1989). What is high school economics? Factors contributing to student achievement and attitudes. *Journal of Economic Education, 20(1)*, 23–38.

Walstad, W. B., Rebeck, K., and MacDonald, R. A. (2010). The effects of financial education on the financial knowledge of high school students. *Journal of Consumer Affairs, Special Issue: Financial Literacy, 44(2)*, 336–357.

Watts, M. (2005). *What works: A review of research on outcomes and effective program delivery in precollege economic education.* New York: Council on Economic Education (formerly the National Council on Economic Education). Retrieved from http://www.councilforeconed.org/wp/wp-content/uploads/2011/11/What-Works-Michael-Watts.pdf.

Way, W., and Holden, K. (2010). *Teachers' background & capacity to teach personal finance.* National Endowment for Financial Education. Retrieved from http://www.nefe.org/what-we-provide/primary-research/grant-studies-teachers-preparedness-and-money-man.aspx.

2

I Wish They Had a Course Like That When I Was in High School

JOHN PELLETIER

Contents

Introduction

Our nation has many high school financial literacy champions. These financial literacy advocates are often the passionate, engaging, and dedicated educators who are providing personal finance training to our future citizens. Despite what some studies may suggest, these hardworking educators appear to be making a difference in the financial lives of many of their students. Witness for yourself, firsthand, the high levels of engagement of high school students with these educators in their classrooms.

Over the past five years, I have had the good fortune to train and partner with many high school educators who are bringing personal finance education into their classrooms. Sometimes this is accomplished via a stand-alone personal finance course that is either a requirement for high school graduation or an elective course. Alternatively, such concepts are imbedded as a significant part of another course offering such as economics or a life skills course.

These educators all have similar stories that they tell about encounters with parents of the children they are teaching personal finance to. Parents often tell them they wished that they had a course such as this when they were in high school. The parents are genuinely grateful for the training being given to their child and often a little envious. The students also proudly report to the teachers that they are providing their parents with new and useful information from the class, such as how credit scores work and how to obtain a free credit report.

Yet, despite these stories, many studies question the efficacy of personal finance education in the classroom. Some researchers claim there is no causal link between financial education and increased financial literacy, better financial behaviors, or improved financial outcomes. They suggest that providing financial literacy education for high school students is terribly misguided. How do we reconcile these disappointing studies and conclusions with the parents' strong endorsement of these classes? One way to resolve this conflict is to focus future research on high school students residing in states or school districts that appear to have the best financial literacy educational conditions in the country.

The focus of future secondary school research should be on student knowledge and behaviors in public high schools in those few states that require an intensive amount of personal finance instruction as a graduation requirement. By focusing on states with the strongest personal finance education requirements in the nation, we will be able to most accurately measure the impact of financial literacy education occurring in our high schools. If personal finance education does not show measurable impacts in student knowledge and behaviors in the few states with the strongest requirements and best program execution, then the rationale given for such personal finance educational efforts becomes quite suspect. Alternatively, if such student educational interventions in these states result in positive and measurable increases in financial literacy knowledge and behavior, then it would appear that this type of intervention can succeed when the right conditions for success exist.

The Case for High School Financial Literacy

Why does it matter if young adults in America have personal financial skills? The reason is quite simple. Financial literacy is crucial to their success in life as adults. It means that they have the ability to use knowledge and skills to manage financial resources effectively for a lifetime of financial well-being.

The need for financial literacy skills for our youth is clear:

- The number of financial decisions an individual has to make continues to increase, and the variety and complexity of financial products continues to grow.
- Many students do not understand that one of the most important financial decisions they will make in their lives is choosing whether they should pursue postsecondary education after high school, and if they decide to pursue additional education what field to specialize in.
- When the majority of college students borrow to finance their education, they often do so without fully understanding how

much debt is appropriate for their education or the connection between their area of study and the income level that they can expect upon graduation. Many students attend college without understanding financial aid, loans, debt, credit, inflation, and budgeting.

- Young people often do not understand the increasingly complex financial products and services they will use as adults (e.g., debit and credit cards, mortgages, banking, investment and insurance products and services, payday lending, rent-to-own, credit reports, and credit scores).
- There has been a large reduction in the offering of pension plans to employees and a large increase in the offering of defined contribution workplace retirement programs. Defined contribution plans impose greater responsibilities on our future workers to save, invest, and spend retirement savings wisely. If they fail to do this, they could become a significant economic burden on our society.
- A recent study indicated that only 24% of Millennials (ages 18 to 34) surveyed could answer four out of five questions correctly in a financial literacy quiz covering fairly basic and fundamental questions (Mottola, 2014). By comparison, 48% of baby boomers (born between 1946 and 1962) were able to answer four out of five of these quiz questions correctly. Clearly, our young citizens are less financially literate than their elders.
- Too few high school students have received personal finance education in school or at home. In fact, a survey indicated that parents are nearly as uncomfortable talking to their children about money as they are discussing sex (Charles Schwab & Co., 2011).

The recent financial crisis clearly shows that a lack of financial literacy was one of the factors contributing to poor financial decisions by individuals and that the choices made had tremendously negative effects on our country.

As former President Bill Clinton recently stated at a conference, financial literacy is "a very fancy term for saying spend it smart, don't blow it, save what you can and know how the economy works (Klein and Giordano, 2014)." Financial literacy, just like reading, writing, and arithmetic, builds "human capital" by empowering individuals with the ability to create "capital for humans" to use in their lifetime—for buying a home, going to college, having a rainy day and a retirement fund. Financial literacy education is not a handout; rather, it is a helping hand that gives individuals the knowledge and skills that can lift them out of a financial problem, or prevent difficulties from occurring.

If financial sophistication is an essential twenty-first century life skill that young people need to succeed, why have our youth failed to master these topics? We would not allow a young person to get in the driver's seat of a car without requiring driver's education, and yet we allow our youth to enter the complex financial world often without any related education. An uneducated individual armed with a credit card, a student loan, and access to a mortgage can be nearly as dangerous to themselves and their community as a person with no training behind the wheel of a car.

The basics of personal financial planning—teaching young people about money; the value of money; how to save, invest, and spend it; and how not to waste it—needs to be taught in school and at home. Without improved financial literacy, the next generation of American leaders, job creators, entrepreneurs, and taxpayers may not have the skills they need to survive and to thrive in this increasingly complex financial world.

When they graduate, high school students should, at a minimum, understand how credit works, how to budget, and how to save and invest. College graduates should understand those concepts in addition to the connection between income and careers, and how student loans work. Adults also need to understand the critical importance of rainy day and retirement funds, and the amounts they will need in those funds.

Why Focus on High School Personal Finance Education?

For research-related purposes, public high school financial literacy programs are an appropriate place to focus time and effort. State financial literacy education requirements and standards tend to be most specific at the high school level and less specific or nonexistent in states' elementary and middle schools. In addition, state-specific personal finance education requirements are tracked by a variety of organizations at the high school level. Much less has been done to accurately identify what is happening in elementary and middle school.

Admittedly, a high school focus could omit students who have dropped out of high school. The National Center for Education Statistics indicates that the high school dropout rate (the percentage of 16- through 24-year-olds who are not enrolled in school and have not earned a high school credential) was 7% in 2012 (U.S. Department of Education, 2015). For many, high school is the beginning of financial independence. In addition, educational opportunities end with high school for many individuals. According to the Bureau of Labor Statistics, only 68.4% of 2014 high school graduates were enrolled in postsecondary institutions (U.S. Department of Labor, 2015).

College financial literacy training data are not being compiled on a national basis, which makes sense given the number and variety of public, nonprofit, and for-profit schools. For those high school graduates who choose to go on to higher education, personal finance education in college is scant and scattered, with few colleges offering a personal finance elective and even fewer requiring personal finance instruction as a graduation requirement. For those who either do not complete high school or choose not to go to college, they will immediately be thrust into a situation where they need to know how to manage their daily living expenses. Given these facts, high school seems like the best and most logical place to deliver personal finance education to America's youth.

Because data are tracked on what is required to be taught on personal finance at the high school level, high schools in states that require substantive personal finance education appear to be fertile ground for testing the efficacy of personal finance education. The state data may also help identify potential control groups. States with no high school personal finance education requirements may serve as useful control group proxies.

In 2013, Champlain College's Center for Financial Literacy graded all 50 states on their efforts to produce financially literate high school graduates (Pelletier, 2013). What the grading shows is that few states have stringent personal finance education requirements. The Center is currently updating its research for a report it will issue in the fall of 2015. Based on our updated research, few states have been identified that required a stand-alone course exclusively focusing on personal finance topics (generally a one-semester, half academic year course) as a high school graduation requirement. In the Center's 2013 study, 40% of states were given grades that you would want your children to bring home from school—grades A or B. Sixty percent of states have grades of C or lower, with 44% having failing grades of D or F. Clearly, you would expect very different results from these states on measurements of student personal finance knowledge and behaviors based on this variance in education efforts.

What Do the Studies Say?

The basic question that researchers in the area of financial literacy education focus on is quite simple: Do financial literacy education interventions lead to better personal finance knowledge and behaviors for individuals as a result of their participation in these programs? Based on one analysis of 188 studies, the results are mixed (Miller, Reichelstein, Salas, and Zia, 2014). From this large group of studies summarized in the report, 18 can be identified that are focused on personal finance education in U.S. high schools. Of those, 10 studies find some positive knowledge and/or behavioral changes of high school students as a result of an educational intervention, whereas eight studies find no material increase in personal finance knowledge from these interventions. However, three of these eight studies identify an increase in savings and certain other financial behaviors after the education intervention. This 56% to 44% ratio of positive to negative financial literacy education study results reminds one of a jump ball in basketball. Based on this analysis, there is no clear and convincing winner in this debate regarding the efficacy of high school personal finance training on student knowledge and behaviors.

Let us review, in a fairly summary fashion, major arguments made with regard to personal finance education at high school and across all ages of our society. This is not intended to be a definitive and comprehensive review of the literature, but merely a summation of some of the major arguments for and against personal finance education efforts with references to some of the research supporting these perspectives.

Financial Literacy Knowledge Leads to Better Personal Finance Behaviors

There are a variety of studies that indicate that individuals with higher levels of financial literacy exhibit more positive personal finance behaviors than individuals with lower levels of financial literacy (Gale and Levine, 2010; Lusardi and Mitchell, 2014). These studies stand for the proposition that individuals who are financially literate appear to make better personal finance decisions. Individuals who are less financially literate are less likely to have a checking account, rainy day emergency fund, retirement plan, and to own stocks. These low literacy individuals are also more likely to use payday loans, pay only the minimum amount owed on their credit cards, have high cost mortgages, and have higher debt and delinquency levels. These types of findings are intuitive. You would expect the most financially literate to make the wisest personal finance choices.

Financial Literacy Education Increases Financial Literacy Knowledge

Other studies support the proposition that financial literacy education interventions can increase financial literacy knowledge (Brown, Grigsby, Klaauw, Wen, and Zafar, 2014). As greater financial literacy appears to lead to better personal finance behaviors, programs that can successfully increase the personal finance knowledge of individuals could be quite beneficial to society. Training programs that increase the number of financially literate citizens should result in an increase in the number of individuals making better and wiser financial decisions in their own lives. However, some studies note that the impact of the personal finance education interventions fade over time.

Financial Literacy Education Does Not Increase Financial Knowledge

Studies supporting the efficacy of financial literacy education efforts suffer from many flaws and weaknesses that make their findings and conclusions suspect (Willis, 2008). In addition, many studies show little or no positive impact from financial literacy education efforts. Financial literacy programs are also costly and time consuming. If it is true that financial literacy education does not work, public policy should not focus on these types of self-help education programs for individuals. Instead, policymakers should focus on the regulation of consumer financial products and services. This approach promotes treating financial products and services much like the Food and Drug Administration treats products it regulates—the goal is to prevent individuals from harming themselves.

Financial Literacy Education Does Not Increase Financial Literacy Knowledge but Appears to Have Positive Impacts on Financial Behaviors

Financial literacy education does not measurably increase personal finance knowledge, but there is evidence that such educational interventions improve certain personal finance behaviors (Mandell and Hanson, 2009). Assessments of financial literacy knowledge may be less important than measurements of behaviors after the completion of the personal finance educational intervention.

Current Financial Literacy Education Efforts Are Not Successful but May Work if We Use a Different Method of Instruction

A meta-analysis covering more than 200 financial literacy education studies finds insignificant increases in financial literacy knowledge and notes that financial education decays over time (Fernandes, Lynch, and Netemeyer, 2013). Based on these results, the authors recommend smaller educational interventions for individuals just before a specific financial decision is made. These "just in time" education efforts may be tied to key events such as buying a car or a home or making an asset allocation decision on a defined contribution retirement plan.

Financial Literacy Education Is Not the Cause of Good Personal Finance Behaviors

Financial literacy knowledge is not necessarily the result of financial education but rather the result of other factors such as mathematics training or individual behavioral traits (Cole and Shastry, 2008; Cole, Paulson, and Shastry, 2014). There is little or no correlation between good financial behaviors and financial literacy education interventions. Instead, these positive financial behaviors appear to be correlated to something other than personal finance educational programs.

Are We Asking the Right Questions and Measuring the Right Things?

Financial literacy efficacy in high school needs to be viewed through a different lens before we can definitively conclude that it is a failure or a success. Here are a few questions that need to be properly addressed by researchers in studies before declaring personal finance education in high schools a failure or a success:

- What are the learning objectives covered by the financial literacy instruction? In some states, the curriculum standards are fairly sparse and vague—even when personal finance instruction is a graduation requirement.
- What curriculum is being used in the classroom to teach this subject? Has the curriculum been proven to work in the classroom by third party experts or studies?
- How many hours of financial literacy instruction are included in the instructional design? If a one semester class has 60 hours of classroom instruction time, what percentage of this time is exclusively allocated to financial literacy topics? There is a significant difference between a stand-alone personal finance course and an economics class that allocates 20% of a one-semester course (approximately 12 hours of instruction) to a smattering of financial literacy topics. We may find that consistent and significant exposure to these topics over time is what is required for success.
- How confident is the educator in his/her ability to teach personal finance?
- Does the educator exhibit good or bad personal finance behaviors in their own personal life?
- Who is allowed to teach personal finance? Does it require a special teaching endorsement, academic background, special training (e.g., a personal finance college or graduate level course)? Or can anyone be asked by an administrator to teach a personal finance course? Most states have fairly stringent requirements on what background an educator must have to teach most high school courses such as mathematics, language arts, and science. Sadly, such requirements are rarely applicable to financial literacy educators. Being self-taught is not optimal for education results.

- What type of standardized assessment will be given to students who take a personal finance course to measure their knowledge? Is this a professionally created end-of-course examination that has been subjected to a well-documented field test and validation study? Were these questions created by assessment question experts or by academic researchers with no psychometric review and expert educator validation?
- Is there funding available for the educator and the school for quality curriculum and teacher continuing education?
- Do educators have easy access to a robust state-sponsored online clearinghouse of vetted and trusted educational resources? This is more than a list of resources. It is a group of resources that are highly recommended by peer educators and is kept current.
- Was a control group of students used who were not exposed to the financial literacy education intervention? Was this a randomized evaluation? Such a control group will allow you to compare data of students who have obtained the financial literacy intervention with students who have not been part of the educational program.
- Is the financial literacy course or instruction required to be taken by all students or is it an elective? There may be self-selection bias issues created by testing an elective requirement rather than a mandate. For example, in some schools, financial literacy instruction is offered as a mathematics-based credit taken primarily by students who do not pursue postsecondary education. This course is taken in lieu of more rigorous mathematics courses such as Algebra II or Pre-Calculus.
- What are the appropriate behavioral metrics to review and measure when doing a longitudinal study on the efficacy of high school personal finance education on students after graduation? Information on credit scores, use of banking services, bankruptcies, credit default rates, and retirement savings would be useful metrics to consider.

Currently, there has not been a series of well-structured studies that takes into account all or even most of the issues addressed in this list.

A Theoretical Framework for Testing the Efficacy of High School Personal Finance

What if we could create a nearly perfect test environment to determine, once and for all, whether personal finance education in high school is a wise public policy solution or a waste of time and effort? What would such a test environment look like? The following is a proposed theoretical framework for testing whether financial literacy education in high school is effective.

1. *Financial literacy education is a high school graduation mandate required to be taught in a stand-alone one semester course.* Research should be focused on states or school districts that require a stand-alone personal finance course as a graduation requirement. Having a one semester course graduation requirement will result in measurements of high intensity personal finance training. A single semester course requirement will be approximately 60 hours of classroom instruction time (based on a Carnegie Unit). This material amount of time excludes time that students will spend on financial literacy homework. It appears that much of the research, historically, has not included this level of instructional intensity. When research focuses on states or schools that incorporate personal finance topics into another course, such as economics, the amount of time and effort placed on financial literacy education may be questionable.
2. *Robust financial literacy education learning objectives and standards.* Research should be focused on states and school districts that have excellent personal finance educational standards and learning objectives. These standards should compare favorably with highly regarded, nationally and/or internationally recognized financial literacy standards. Three source documents to use when judging these educational standards should include: (1) Jump$tart Coalition for Personal Financial Literacy's fourth edition (2015) of the *National Standards in K–12 Personal Finance Education*; (2) Council for Economic Education's 2013 *National Standards for Financial Literacy*; and (3) Organization for Economic Co-operation and Development's (OECD) Program for

International Student Assessment's (PISA) 2013 publication *PISA 2012 Assessment and Analytical Framework; Mathematics, Reading, Science, Problem Solving and Financial Literacy.*

3. *Clear understanding of the subject matter expertise of educators delivering the training.* Data should be gathered in a manner that will allow researchers to determine whether certain methods of personal finance education delivery work better than others. For example, a state may allow a financial literacy mandate to be taught by a variety of educators. Their technical expertise could be in business, social studies, mathematics, technology, agriculture, family and consumer sciences, and even drivers' education. It is imperative that we understand the background of the educators involved. It is possible that measurements of student knowledge and behaviors could be dramatically different based on this single factor. The subject matter specialty of educators should be an important data point to capture.
4. *Determine whether there are multiple pathways to meet the mandate and measure pathway differences.* Is a financial literacy course offered through a variety of options at each high school? Some high schools that require personal finance as a graduation requirement offer it as both a social studies course and as a mathematics course. The course offerings are designed to meet two distinct graduation requirements. For example, if a high school mandates four mathematics courses as a graduation requirement and three social studies courses as a graduation requirement, a financial literacy course is available that will fulfill either requirement. Schools that meet a mandate in this fashion may have very different students self-select the type of course offered. In certain high schools, the students who typically do not go to postsecondary institutions after high school meet the financial literacy mandate through a mathematics credit, whereas those planning to attend a postsecondary institution after high school often meet the mandate through a social studies elective requirement. This type of self-directed segregation may, or may not, be an import factor in the overall student knowledge and behavior results and should be tracked if possible.

5. *Proven quality curriculum and tools and an online educator resource tool available.* Research should rate the quality of the curriculum and other tools used by the educators in the state or school district when implementing the financial literacy mandated course. Is a nationally recognized curriculum being used in the classroom instruction? Has it been tested for its efficacy? Does it include the educational topics and learning objectives outlined in nationally recognized educational standards prototypes for financial literacy (see item number 2)? Do the educators have access to a financial educator website with trusted and vetted recommendations for classroom tools that are available for instructional support (grade-appropriate curriculum, lesson plans, videos, games, activities, projects, case studies, calculators, books, articles, field trip opportunities, etc.)? Has the online resource been created with input from financial literacy educators? The educator must have access to quality curriculum and classroom tools to be successful.
6. *Requirements on who is allowed to teach personal finance and continuing education expectations.* Research should be done in a state or school district that requires educators who teach the mandatory personal finance course to have a minimum level of financial literacy training. This can be done through the requirement of an educator endorsement as well as a continuing education requirement tied to the renewal of the endorsement. An endorsement is a statement appearing on an educator's teaching license that identifies specific subjects and/or grade level that the license holder is authorized to teach. The supermajorities of states do not have endorsement or continuing education requirements related to the teaching of personal finance course in high school.

In 2011, the Center for Financial Literacy at Champlain College ("Center") conducted a survey regarding high school personal finance education that was sent to all Vermont administrators (principals, assistant principals, superintendents, assistant superintendents, and curriculum directors) (Center, 2011). An average of 95% of respondents felt that they had teachers currently on staff who could teach specific financial literacy standards (based on the Jump$tart Coalition educational standards). Ironically, these same administrators concluded that there was a woeful lack of personal finance instruction available for educators. Having someone who "could" possibly teach a financial literacy course is very different from having a highly trained educator teach a personal finance course. Financial literacy high school educators come from many different subject matter areas of expertise. An educator who is keenly interested in financial literacy topics may have never received professional training on how to teach a personal finance course. Just imagine the reaction of parents if untrained educators were allowed to teach language arts, mathematics, history, civics, or a foreign language.

Teacher training in financial literacy is critical to educators' confidence—and likely their ability to be effective, yet it is often not required and may not even be readily available for educators. The Center's high and middle school graduate level financial literacy educator training program (consisting of 45 hours of instruction) participated in a study that measured the impact of similar, but not identical, educator training programs conducted by different groups in three states. Each site provided robust educator training on personal finance topics and through surveys obtained information on educator confidence, attitudes, and behaviors before and after the education intervention. The percentage of participants who agreed they had the knowledge necessary to effectively teach their students about personal finance increased from 38% before the intervention to 80% after the intervention (Hensley, 2013). After the education was completed, the educators made positive financial changes in their own lives. For example, the percentage of participants who took steps to improve their credit scores increased from 39% before the training to 71% after the training. This study shows how impactful educator training in this area can be.

7. *Measure educators' attitudes, personal behaviors, and confidence levels.* A successful study should not only measure the results of the students' behavior and knowledge, but it should also take into account the attitudes, personal behaviors, and confidence levels of the educators involved in the program. Are

there correlations between educators' attitudes and personal behaviors and the knowledge and behavior results of their students? Do correlations exist between an educator's confidence in his/her ability to teach personal finance concepts in the classroom and the results of the students taught by the educator? Interesting insights could be obtained if it were possible to capture such educator data in a confidential manner.

A 2009 survey of more than 1200 K–12 teachers, students in college learning to be educators, and their professors sought to better understand participants' training and education in personal finance, their opinions about the importance of financial education, and their capacity to teach these topics. Here are some of the findings of this study (Way and Holden, 2009):

- Nearly 64% of teachers did not feel qualified to use their state's financial literacy standards.
- Fewer than 20% of teachers reported feeling "very competent" to teach any of the six personal finance topics surveyed.
- Only 37% of K–12 teachers had taken a college course offering financial education-related content (this could have been an economics course).
- Teachers who had taken a college course with financial education-related topics were 50% more likely to rate themselves as competent to teach financial literacy subject matter.
- Nine out of ten teachers believe a personal finance course should be a high school graduation requirement.

Studies measuring an educational program with a small test group of educators often lack any data or insight into the educators themselves and whether they are prepared from a training, confidence, or life experience perspective to effectively teach a financial literacy course. In the future, capture of this information would be useful, if it is possible to obtain.

8. *Use of standard assessment test created by experts that has been field tested and validated.* Along with mathematics, reading, and science, financial literacy is now considered an essential skill that young people need to succeed in life. In fact, the OECD's PISA recently added financial literacy to the topics

it measures across countries in a test given to 15-year-olds. The results of this international assessment are depressing. The United States ranked ninth out of 13 countries that participated, behind many current or former communist countries and in a statistical tie with the Russian Federation (OECD, 2014). Despite this international focus, the United States does not have a proven and generally accepted critical skills-based financial literacy assessment tool for high school students. Assessment tests, when they exist at all, tend to be curriculum-based. Curriculum-based assessment is not consistent with how educators measure other subjects, such as language arts and mathematics.

Researchers should use a broad-reaching and effective financial literacy assessment test to measure whether education efforts lead to greater knowledge. An effective test will measure the full range of knowledge and skills identified as critical to being an effective participant in the economy without relying on any specific curriculum or instructional model. Although there have been a number of efforts to develop financial literacy assessments, they suffer from several serious weaknesses. As mentioned earlier, existing financial literacy tests are largely curriculum-based and oftentimes measure a narrow, idiosyncratic curriculum. In addition, financial literacy knowledge assessments have little consistency from one study to another. And, given the difficulty of creating high-quality assessments, these assessments are often technically flawed and do not accurately measure the financial literacy skills needed.

Researchers should have access to a quality test based on a set of standards that incorporates topics addressed in seminal financial literacy literature and widely accepted national, state, and international financial literacy programs. This means that students in different states or school districts who take the assessment will be tested on content standards that are critical to success, regardless of what curriculum was used to teach them.

All questions in such a test should be reviewed by psychometricians (experts who measure the validity, reliability, and fairness of an examination program) to address psychometric issues. An equity reviewer should review each question prior to release to ensure representativeness and freedom from bias. Other review concerns that should be addressed are: (1) clarity—all questions should be presented in a straightforward manner and readability should be managed to ensure that it does not interfere with the examinees ability to respond; (2) importance—all questions should reflect important content and skills; content and skills that are considered important and are commonly taught; and (3) evidence based—each question is in essence a source of evidence about students' knowledge or skill levels; each item should clearly provide evidence of the skill and knowledge claims made. Psychometric analysis of these materials is critical not only to ensure that the items and assessments are of high quality but also to ensure that the scores yielded by these assessments are clearly interpretable. Researchers should have access to a test, created by assessment experts, that has been subjected to a field test and validation study.

Instead of a standard assessment tool, authors of financial literacy studies often create their own idiosyncratic set of assessment questions. If the student assessment tool is flawed, then the results derived from the assessment will be flawed as well.

Researchers and policy makers need high-quality data on students' financial literacy skill levels in order to make informed decisions on how to structure successful financial education strategies in high school. Data are needed to (1) expose gaps in financial literacy knowledge; (2) identify which financial literacy education strategies are the most successful; (3) find best practices in the classroom that can be shared with educators across our nation; and (4) come up with ongoing efficiency improvements in personal finance education. An effective test will help ensure that financial literacy standards are being met and will identify topics and standards that require further emphasis. A standard test will allow for the measurement of growth in financial literacy—for students, classrooms, schools, geographies, and other relevant subgroups. A robust measure of financial literacy among young people will provide information at a national level that can indicate whether the current approach to financial education is effective. A standardized professionally created test will be a baseline from which to measure success and review programs in future years. The creation of a well-regarded national financial literacy assessment tool should be funded and made available to states, school districts, and researchers.

9. *Funded educational mandate.* Research should be focused on a state or school district that is willing to fund robust financial literacy education program in high schools. Such a state or school district will allocate monetary resources to schools and educators so that they can have access to quality curriculum, classroom tools, educator training, and a quality end-of-course assessment tool. Unfunded mandates are not likely to be successful from an implementation perspective.
10. *Use of a high school control group.* Researchers, when conducting a study, should also measure the knowledge and behavior of a group of high school students who have not taken a financial literacy course. These "control group" students should be similar in age and other demographics to the students who have been exposed to the mandated personal finance course. Most high school studies only focus on the impact of a course or of specific curriculum on the students who have participated in the specific intervention. Consistent with the best scientific research standards, the students who are part of the "treatment" group obtaining the mandated training should be compared to a "control group" that is not.
11. *Longitudinal study and key factors measured.* Longitudinal research should be conducted following students who have taken a stand-alone personal finance course that was a graduation requirement. Do these students who have been required to take an intensive personal finance course have better financial outcomes than students who did not take such a course in high school? Given the reality that knowledge gained from any high school course is likely to fade over time, longitudinal studies should follow students immediately after graduation from high school for at least one decade. Rather than measuring knowledge-based questions (e.g., Who was the fifth president of the United States?), perhaps we should focus on

behavior measurements. When controlling for certain factors (e.g., educational attainment and income), do students who are required to take a personal finance course exhibit better financial behaviors than individuals who have not had access to a personal finance course as they enter adulthood? Some measurements could include credit scores; use of checking and savings accounts; credit card payment habits; default rates on credit (mortgage, credit card, auto loans, and student loans); ownership of financial assets such as mutual funds, stocks and bonds; retirement plan participation and savings for retirement and an emergency fund. Some of these data are currently available at the national, state, and zip code levels.

Recent High School Studies That Meet Some of the Proposed Framework Standards

Two recent studies meet some of the study framework requirements and both have fairly impressive results as they relate to the efficacy of high school personal finance education.

- *Financial literacy mandates improve credit behavior.* One group of researchers focused on three states where material personal finance high school education mandates were recently enacted (Brown, Collins, Schmeiser, and Urban, 2014). Using robust data, they measured the default rates and credit scores of recently graduated students who were subject to these new mandates and compared them to the results of similarly aged individuals in nearby states that did not have financial literacy high school mandates. The researchers concluded that mandatory personal finance education in high school improves the credit behavior of young adults, despite the fact that in each state, the personal finance education was delivered as part of an economics class requirement.
- *Robust educator training and well-designed curriculum works.* Another study shows that a well-designed personal finance course (one semester in length) taught by highly trained educators who attended a 30-hour weeklong training program and who used specific curriculum improved the average personal

finance knowledge of the students in all standard and concept areas covered by the researchers' assessment examination (Asarta, Hill, and Meszaros, 2014).

Do We Hold Personal Finance Courses to Higher Standards than Other High School Courses?

We need to have reasonable expectations with regard to what a single high school course can achieve over the lifetime of an individual. Perhaps academic researchers expect too much from a single high school course taken by students with varying degrees of personal finance education. We expect that someone who has had some level of personal finance instruction in high school for one semester, often as part of another course offering such as economics (given for one-half of an academic year), should have substantially better financial literacy results than someone who has never taken that course when they reach their thirties, forties, or fifties. Remember, less than a handful of states mandate a stand-alone personal finance course as a graduation requirement. Those high school students who are required to take a course—such as economics, civics, or a life skills course that is imbedded with financial literacy topics—may be getting minimal classroom instruction on these topics. These allegedly better-trained individuals are expected to have substantially better results on very specific personal finance knowledge questions. They must also have better credit scores, savings, more retirement savings, fewer bankruptcies, and lower credit defaults and credit card debt much later in life. Decades later, this single personal finance educational experience (which in some states could be less than 10 hours of instruction) must be proven to have materially changed their lives or we must conclude such education is a woeful failure.

Yet we do not hold other high school courses focused on a single topic to this exacting requirement. Do we hold civics, foreign languages, and calculus to these high knowledge and behavior standards?

We know from a variety of studies in the area of personal finance education that the effect of this type of education tends to fade away over time as individuals age. Not surprisingly, a course taken 5, 10, 20, or 30 years ago can lose its impact. Yet, the same could probably be said about virtually any other high school course.

Civics is the study of rights and duties of citizenship. Civic topics are required to be taught in many high schools. One basic duty is voting, particularly for the U.S. president. Voter turnout in presidential elections since 1972 has been consistently well below 60% (The American Presidency Project, 2015). Based on this result, should we stop teaching civics in the classroom? Of course making such an argument would be silly. Just because we do not get the voter turnout percentages we desire does not mean we should stop this form of education or that this education is not useful.

Imagine if the quality of a high school calculus class was judged solely by the ability of 30-, 40-, or 50-year-olds to pass a standardized calculus assessment examination such as the College Board's Advance Placement Calculus test decades after taking the course in high school? Needless to say, even a calculus class taught by an exemplary teacher would not fare well based solely on results 10–30 years later. We expect that personal finance knowledge should survive without diminution for decades but would never expect that from any other topic of learning in high school.

Education is cumulative. With mathematics you start with counting, move on to addition and subtraction and then to division and multiplication. You move slowly from algebra to calculus. You learn letters before you can read.

Yet for some reason, we do not apply cumulative learning concepts to our understanding of the efficacy of personal finance education. Most of the studies measure a single personal finance intervention in high school or as an adult, and many of the adult studies are focused exclusively on low-income adults.

In a perfect world, we could measure the success or failure of a truly cumulative approach to financial literacy. This would require that the topics be incorporated, in an age-appropriate fashion, from grades K through 12 and even into the collegiate setting. Clearly, this level of consistent financial literacy educational engagement is hard to find anywhere in the country. Often educators lack training on how to incorporate personal finance into the classroom. Where teacher training exists, it is usually focused on high school teachers. Middle and elementary school teachers are often an afterthought. Even when a state requires personal finance concepts from grades K–12, the state often does not keep track of or measure how this is actually being accomplished at the local school district level. And let us not forget the scarcity of educator training dollars for financial literacy and the limited educational opportunities available for many teachers in this subject area.

A K–12 cumulative financial literacy program administered by highly trained professionals has never been developed, let alone studied and measured. Yet researchers seem to expect similar knowledge and behavior results that are comparable to K–12 cumulative training programs that exist in other subject matter areas such as mathematics and in language arts. Is this even a realistic or fair way to judge the efficacy of personal finance education?

Conclusion

The appropriateness and effectiveness of high school personal finance educational initiatives must be measured. However, such measurement should take place under test conditions where they are most likely to succeed. Let us focus on identifying the right states and school districts to test. We need to understand the intensity of the education program; the quality of the educators and their subject matter expertise; the quality of the curriculum and teaching tools used; the quality of the assessment process given to students; and a longitudinal study of key behavioral metrics before concluding that financial literacy education is a success or a failure. We also need to have a control group of students who are not taking the course. If the instruction is poorly executed when designed and when taught, then you would expect poor student results from the end-of-course assessment and after the completion of the course in a longitudinal study. Only when we have the right test conditions will we truly know whether high school financial literacy educational interventions are useful or a waste of time, effort, and resources.

References

Asarta, Carlos, Andrew Hill and Bonnie Meszaros. 2014. The Features and Effectiveness of the Keys to Financial Success Curriculum. *International Review of Economics Education* 16: 39–50. Retrieved from http://www.sciencedirect.com/science/article/pii/S1477388014000140#.

Brown, Alexandra, J. Michael Collins, Maximilian Schmeiser and Carly Urban. 2014. State Mandated Financial Education and the Credit Behavior of Young Adults. Finance and Economics Discussion Series Divisions of Research & Statistics and Monetary Affairs Federal Reserve Board Working Paper 2014-68. Retrieved from http://www.federalreserve.gov/pubs/feds/2014/201468/201468pap.pdf.

Brown, Meta, John Grigsby, Wilbert van der Klaauw, Jaya Wen and Basit Zafar. 2014. Financial Education and the Debt Behavior of the Young. Federal Reserve Bank of New York. Staff Report No. 634. Retrieved from http://www.newyorkfed.org/research/staff_reports/sr634.pdf.

Center for Financial Literacy, Champlain College. 2011. Financial Literacy Education in Vermont High Schools: A Snapshot. Retrieved from http://www.champlain.edu/centers-of-excellence/center-for-financial-literacy/cfl-resources/financial-literacy-education-in-vermont-high-schools.

Charles Schwab & Co. 2011. 2011 Teens & Money Survey Findings, Insights into Money Attitudes, Behaviors and Expectations of 16- to 18-Year Olds. Retrieved from http://www.schwabmoneywise.com/public/file/P-4192268/110526-SCHWAB-TEENSMONEY.pdf.

Cole, Shawn and Gauri Kartini Shastry. 2008. If You Are So Smart, Why Aren't You Rich? The Effects of Education, Financial Literacy and Cognitive Ability on Financial Market Participation. Paper presented at the 2009 Federal Reserve System Community Affairs Research Conference, Washington, D.C.

Cole, Shawn, Anna Paulson and Gauri Kartini Shastry. 2014. High School Curriculum and Financial Outcomes: The Impact of Mandated Personal Finance and Mathematics Courses. Harvard Business School Working Paper 13-064. Retrieved from http://www.hbs.edu/faculty/Publication%20Files/13-064_c7b52fa0-1242-4420-b9b6-73d32c639826.pdf.

Fernandes, Daniel, John Lynch, Jr. and Richard Netemeyer. 2013. The Effect of Financial Literacy and Financial Education on Downstream Financial Behaviors. National Endowment for Financial Education. Retrieved from http://www.nefe.org/Portals/0/WhatWeProvide/PrimaryResearch/PDF/CU%20Final%20Report.pdf.

Gale, William and Ruth Levine. 2010. Financial Literacy: What Works? How Could It Be More Effective? Brookings Institution. Retrieved from http://www.brookings.edu/~/media/research/files/papers/2010/10/financial-literacy-gale-levine/10_financial_literacy_gale_levine.pdf.

Hensley, Billy. 2013. Research Report, Content-Based Teacher Professional Development Pilot Project. National Endowment for Financial Education and Jumpstart Teacher Training Alliance. Retrieved from http://www.jumpstart.org/assets/files/Teacher%20Alliance/J$TTA%20Pilot%20Research%20Report.pdf.

Klein, Asher and Jackie Giordano. 2014. Bill Clinton Visits USC to Teach Kids Value of Financial Literacy. Channel 4, Southern California. Retrieved from http://www.nbclosangeles.com/news/local/Bill-Clinton-Visits-USC-to-Host-Financial-Literacy-Event-282070241.html.

Lusardi, Annamarie and Olivia Mitchell. 2014. The Economic Importance of Financial Literacy: Theory and Evidence. *Journal of Economic Literature* 52(1): 5–44.

Mandell, Lewis and Kermit Hanson. 2009. The Impact of Financial Education in High School and College on Financial Literacy and Subsequent Financial Decision Making. Presented at the American Economic Association Meetings in San Francisco. Retrieved from https://www.aeaweb.org/assa/2009/retrieve.php?pdfid=507.

Miller, Margaret, Julia Reichelstein, Christian Salas and Bilal Zia. 2014. Can You Help Someone Become Financially Capable? A Meta-Analysis of the Literature. Background Paper to the 2014 Global Financial Development Report. Policy Research Working Paper 6745, World Bank.

Mottola, Gary. 2014. The Financial Capability of Young Adults—A Generational View. *FINRA Foundation Financial Capability Insights*. Retrieved from http://www.usfinancialcapability.org/downloads/FinancialCapabilityofYoungAdults.pdf.

Organization of Economic Co-operation and Development (OECD). 2014. PISA 2012 Results: Students and Money: Financial Literacy Skills for the 21st Century (Volume VI). PISA, OECD Publishing. http://dx.doi.org/10.1787/9789264208094-en.

Pelletier, John. 2013. 2013 National Report Card on State Efforts to Improve Financial Literacy in High Schools. Center for Financial Literacy at Champlain College. Retrieved from http://www.champlain.edu/Documents/Centers-of-Excellence/Center-for-Financial-Literacy/National-Report-Card-Champlain-College-CFL.pdf.

The American Presidency Project. 2015. Voter Turnout in Presidential Elections 1828–2012. Retrieved from http://www.presidency.ucsb.edu/data/turnout.php.

U.S. Department of Education, National Center for Education Statistics and the Institute of Education Sciences. 2015. Fast Facts, Dropout Rates. Retrieved from https://nces.ed.gov/fastfacts/display.asp?id=16.

U.S. Department of Labor, Bureau of Labor Statistics. 2015. Economic News Release, College Enrollment and Work Activity of 2014 High School Graduates. Retrieved from http://www.bls.gov/news.release/hsgec.nr0.htm.

Way, Wendy and Karen Holden. 2009. Teachers' Background and Capacity to Teach Personal Finance: Results of a National Study. National Endowment for Financial Education. Retrieved from http://www.nefe.org/Portals/0/WhatWeProvide/PrimaryResearch/PDF/TNTSalon_FinalReport.pdf.

Willis, Lauren E. 2008. Against Financial Literacy Education. *Iowa Law Review* 94(1): 197–285.

3

How Twentysomethings Are Changing Financial Literacy Education

RAKHEE JAIN

Contents

If you told me when I was 14 years old that it was better to buy a single stock rather than a stock mutual fund in order to minimize riskiness, I would tell you that you were wrong (Rephrasing of Question 3 in the 2009 Financial Capability Study; Lusardi and Mitchell, 2011). The best way to minimize risk is to diversify your holdings. Unfortunately, only about half of Americans will answer this question correctly, and 38.7% of Americans will answer "don't know" (Lusardi and Mitchell, 2011). This is likely because half of Americans do not have fathers such as mine.

My dad has always been of the mentality that your money should work for you. Consequently, I bought my first stock at the age of 9 and was encouraged to save and keep track of spending from a young age. My father is a certified public accountant; he has always been on top of my family's financial situations. Before entering the University of Chicago (UChicago) for my undergraduate studies, my Federal Application for Student Aid (FAFSA) was forwarded directly to my father; I did not even bother to open the email. The forms UChicago uses to determine financial aid were also forwarded straight to my father. In short, when it came to planning for college and figuring out how to finance it, I let my father take the reins while I remained woefully ignorant.

It was with some level of awareness regarding the privilege that came from my financially literate father that kindled my interest in Moneythink in the autumn of my first year as a university student. As a freshman, in a new city, far from home, I was unsure of who I was and what I wanted to get involved with at UChicago. It was at a club fair that I first encountered Moneythink. With its bright orange lightbulb logo and a mission that resonated with me, I joined the organization. Since then, Moneythink has arguably become the most transformative experience for me as an undergraduate.

Simply put, Moneythink is a national nonprofit organization that aims to create economic opportunity through teaching a financial literacy curriculum. Various chapters exist around the country and use the same approach to financial literacy education. The organization relies on college-age mentors to go into the classroom and teach a lesson regarding financial life skills each week (Our Approach, Moneythink National). The consistency of teaching leads to natural and impactful relationships between high school students and their college mentors. Second, the closeness in age between mentor and mentee allows for a relationship to organically form. What we experience in college and the financial decisions we, as university students, face are in the relatively near future for our high school students, who are generally in their junior year of study.

The schools Moneythink operates in are unique. Because financial literacy is most important for minority and lower income students, our goal is to teach in schools where at least 80% of students have free or reduced price lunch, a common indicator of income level.* The UChicago chapter of Moneythink teaches in various types of schools such as Cristo Rey Charter schools and Noble Charter schools, as well as other models of education. We teach in classrooms during the school day as well as afterschool programs. The range of programming and school model allow us to determine what methods of teaching are

* Conversation with Joe Duran, Moneythink chief operating officer.

most impactful and effective for our individual chapter. For example, our students at the Cristo Rey model of school are usually able to hold more complex conversations regarding financial decision making. This is because the Cristo Rey model of charter schools is a work study model in which students pay their tuition through a job they hold. These working students already have debit cards and are generally aware of how credit cards work. With our students in this model of schooling, we usually complete the basic financial literacy curriculum early on and are then able to focus on more complex topics such as the stock market and entrepreneurship. This is different from one of the UChicago charter schools that we teach in, where Moneythink is an afterschool activity that counts as extra credit for math class. A different level of background knowledge exists among these students compared to that in the Cristo Rey model, but they still are very eager to learn and are exceptionally bright high school students.

The history of Moneythink is also one that is striking and important in order to frame much of what I have to say. Moneythink began as a single chapter at the UChicago and has since grown rapidly. Now in more than 30 universities nationally (*History*, n.d.), the organization has grown aggressively since the nonprofit started as a club at the UChicago in 2008.* The most important takeaway is that the organization began at UChicago. Consequently, as a member of the flagship chapter, I have been able to participate and engage with the National team in a number of unique opportunities. Furthermore, the organization itself was created to address a very clear need for financial literacy education within south side Chicago neighborhoods—the same neighborhoods that we, as the UChicago chapter, still serve (Mint, 2015). It is my hope that after reading this chapter, it will be very clear what both Moneythink UChicago and Moneythink, at the national level, have been doing, how effective we have been, and what we hope to accomplish over the next few years.

Before I explore the financial literacy education landscape a little more, I would like to preface all that I have to say with a very strong disclaimer. I am an undergraduate student who has taught with

* The organization's CEO is Ted Gonder, who was recently picked as one of *Fortune Magazine*'s 30 under 30. In short, he is a man making waves while also being an all-around incredible human being.

Moneythink for 3 years. My personal relationship with the National Moneythink team is robust, as I have been fortunate enough to work closely with them. Nonetheless, I am not a member of the National team nor am I an academic with my own personal research regarding financial literacy education within the United States. Much of what I have to say may only be relevant within a very specific framework. Furthermore, my experiences are not reflective of the experiences of every Moneythink mentor.

I am, however, exceptionally passionate about this topic. I hope that what I have observed and the insights from the Moneythink curriculum can help further the discussion and bring about tangible change with what we teach in our schools. It is my hope that one day every 14-year-old will understand the importance of risk diversification. Moreover, I hope that—with better financial literacy education—everyone will make smarter and better decisions regarding their money.

Demographics: Who We Are Teaching and Why

It is true that with higher education comes more knowledge regarding financial decision making. However, surprisingly high levels of financial awareness persist even after significant postgraduate studies. In fact, among young Americans (age range, 25–34 years) with postgraduate education, only 60% of the respondents were capable of correctly answering three basic questions used to determine financial literacy (de Bassa Scheresberg, 2013).

More importantly, the significance of financial literacy cannot be understated. When it comes to long-term financial planning such as retirement planning and emergency savings, those with more financial literacy knowledge are more likely to plan ahead (de Bassa Scheresberg, 2013). High cost borrowing methods such as payday lenders or title loans are also much less likely to be accessed with increasing knowledge of financial literacy (de Bassa Scheresberg, 2013).

Heartbreakingly, it is women and minorities that demonstrate the least knowledge when it comes to financial literacy. And yet, we see the role of women in household financial decision making continuously increasing (Woodyard and Robb, 2012). Furthermore, minority communities are placed at a higher risk of making hazardous financial choices (Woodyard and Robb, 2012). The unfortunate reality is that all of these questionable choices regarding finances can be simply avoided with knowledge.

But, overwhelmingly, we see that the most important groups that must be served in terms of financial literacy knowledge are students and young adults. Only 17 states in the United States require some sort of personal finance course in K–12 education (Economic and Personal Finance Education in Our Nation's Schools 2014, Council for Economic Education). Moreover, we see an alarming trend in debt acquisition. Nearly 20% of students graduated college with more than $7000 in credit card debt in 2009 (de Bassa Scheresberg, 2013). The average student loan debt after gaining a bachelor's degree in 2011 was $26,600. Younger generations are increasingly more likely to file for bankruptcy, and the rate of bankruptcy is highest among the younger age groups (de Bassa Scheresberg, 2013). Clearly, our youth are not being taught how to most effectively manage their money.

We can agree that financial literacy is important, and we see that much of the content of financial literacy concepts is not intuitive, given some people's inability to answer basic questions even with significant education. How then do we teach the next generations what strategies to use when managing personal finances? This is where Moneythink comes in.

The landscape for financial decision making has changed dramatically over the past years. With the change from defined benefit to defined contribution pension plans, the scope of responsibility regarding an individual's financial situation has greatly increased (de Bassa Scheresberg, 2013). This is why we target high school students. High school students are on the cusp of the rest of their lives. For many, it means an undergraduate education or adulthood. Moneythink focuses on high school juniors so that we may target and reach this vulnerable demographic. For many college-bound students, junior year is a pivotal moment in their educational careers as it is the time where they begin to make decisions regarding their higher education. This is also when we would like them to start thinking about becoming financially independent and aware.

Moneythink as an organization has an end goal. It is not the organization's hope to survive. In fact, chief executive officer (CEO) and cofounder Ted Gonder's vision is to be an obsolete organization by 2030 (Carpenter, 2014). That is, in 15 years, financial literacy education will be a normalized and integral part of every high school curriculum. In 15 years, we hope that every child is financially literate and that Moneythink's logo of a bright orange lightbulb will be a fading memory.

Curriculum: What We Teach and Why

The Financial Capability Curriculum (FCC) consists of 12 distinct lesson plans that are meant to be taught within the framework of small groups. Moneythink National has set up four key pillars encompassed within the Moneythink mission. Known as the "impact pillars," the goals of spending mindfully, saving money, making money, and using financial products safely are introduced and taught within the FCC and over the course of the 12 lessons.*

The FCC's success is contingent on the way in which the lessons are taught. Specifically, each mentor works with the same small group of students every week to allow for meaningful connections to be forged. Our goal is to maintain a 5:1 students/mentor ratio. This high touch mentoring model is more effective than lecturing to a classroom of students and allows mentors to get to know their students on a personal level. Not only is the material better communicated, students have a meaningful way to engage with it as well.

Each lesson is designed in a way that allows for introduction to new material, discussion among the group, and time to internalize information through handouts that are completed in groups or individually. An example of a handout used in the first FCC lesson is also included in the following discussion. The entire FCC is available online for free through Moneythink National and is a great way to really understand the impact of the high touch mentoring model and the kinds of lessons we teach in the classroom.

The curriculum is best described in two different parts. The first part focuses on knowledge surrounding basic financial life skills. Topics for lessons vary; financial goal setting, the banking system, and credit scores are all examples of lessons that we teach. The second part of

* 2014 Financial Capability Curriculum Lesson 1, Moneythink.

the curriculum focuses on college readiness and preparing for the working world. Thus, topics such as resume development, interviewing skills, and professional dress are covered in the second part of the curriculum.

The lessons typically address an impact pillar through the content that is taught. The full curriculum list with their associated objectives is included in the following discussion.

Figure 3.1 shows the FCC overview that the Moneythink UChicago chapter uses for personal distribution.

As part of our curriculum, we use various handouts to reinforce the lessons. Figure 3.2 shows an example of a handout we use in the classroom.

Moneythink's Financial Capability Curriculum

Session	Class objectives
Why Moneythink?	Students will be able to identify the two components of financial success: make money and manage money.
Mindful spending	Students will be able to identify impulse spending and understand what it means to spend mindfully.
Goal setting	Students will set ambitious, yet feasible goals for themselves in academic, financial, and extracurricular spheres.
Introduction to banking system	Students will be able to describe the differences between, and pros/cons of, various money habits such as storing money at home and using a prepaid card.
Banking + NerdWallet	Students will be able to identify the steps in evaluating, selecting, and opening a bank account.
Financial product basics	Students will be able to identify the difference between an account number and a routing number, write a check, and highlight the differences between cash and check.
Banking safety 101	Students will be able to communicate detrimental actions within the banking system and how to avoid making common mistakes; students will be able to recite key rules-of-thumb.
Time value of money + regular saving	Students will understand the value of regularly saving and understand the compounding effects of saving early.
Credit	Students will be able to define key rules of thumb for using credit cards and how to build a good credit score.
Marketing yourself	Students will be able to demonstrate comprehension of basic professional dress and behavior.
Resume workshop	Students will be able to construct and revise a professional resume
The world of work	Students will demostrate comprehension of the professional interview process.

Figure 3.1 This curriculum outline serves as a "one pager" that Moneythink UChicago uses to explain the curriculum to current or potential school partners.

Moneythink Financial Capability Curriculum
Lesson 1 – Handout A

Name:
Date:

What Do We Know?

Directions: For each of the following statements, you and your group members will rate the statement on a scale of 1-5. If you are giving the statement a 1, then you believe it is absolutely false. If you are giving the statement a 5, then you believe it to be absolutely true. Choose any number in between if you're not certain.

- You are spending $5 a day on lattes at Starbucks. However, the Student Center at your university has free coffee and milk for you to grab before class.
  This is an example of mindful spending.
  1 2 3 4 5
- Your parents give you a $10 allowance each week, and you usually put half of that money into your piggy bank, and half into your wallet. Your best friend wants to go shopping for new Nike sneakers that cost $120. You decide to take all the money out of your piggy bank and buy the sneakers your best friend bought.
  This is an example of mindful spending.
  1 2 3 4 5
- You have decided to start babysitting for two hours a week, and you earn $5 an hour. You decide that you would like to start saving some of your money. You put half of your earnings into your wallet, and half into your piggy bank.
  This is an example of saving wisely.
  1 2 3 4 5
- A couple of your friends decided that they were going to open a bank account. There is a bank two blocks away from your school. You don't know anything about this bank, but your friends decide that you should open your account here, since it is so close to school. You agree, and decide to open yours there too.
  This is an example of weighing your options wisely and carefully.
  1 2 3 4 5
- You are creating a resume for your first job as a counselor at the park recreation center. You decide to write your resume on a piece of paper that you ripped out of your notebook. You are writing in your favorite purple pen, and are listing all of the things that make you unique and fantastic.
  This is an example of a professional resume.
  1 2 3 4 5
- You have an interview at your local elementary school's summer camp. You decide that you are going to wear your jean shorts, tank top, and flip-flops. Your interview is for 3 PM, but you do not show up until 3:04 PM.
  This is a professional way to conduct an interview.
  1 2 3 4 5

Figure 3.2 Lessons are meant to explore and understand the four impact pillars of Moneythink. Handouts are used to reinforce the main content of our lessons and may be for group or individual work.

Innovation: What Is Mobile and Why It Is Changing the Financial Literacy Landscape

A discussion regarding the FCC would be incomplete without introducing the most cutting-edge innovation and curriculum addendum to date. Moneythink Mobile, a smartphone application, was launched in the 2014–2015 academic year and is already transforming the way mentors teach. Each of the 12 lessons in the FCC is accompanied by a mobile component.

Each lesson is supplemented by an additional "challenge" that is meant to be completed using the mobile phone application. Thus, the material taught in the classroom is reinforced throughout the week until our next meeting with our students. Furthermore, we hope that the phone application will reinforce smart money habits and, eventually, even lead to behavior changes.

The idea behind mobile is straightforward. The only time in which our students are not making financial choices is the time we spend with them in the classroom. Moneythink Mobile's aim is to be present even when mentors are not physically there. Mentors and other students are able to engage with one another by posting, commenting, and liking posts. CEO Ted Gonder refers to the mobile phone application as "an Instagram for finances" (Our Program, Moneythink website).

The Moneythink National team worked together with IDEO.org to design a challenge-based app for students.* These challenges engage our students in a number of ways, requiring them to complete certain tasks and think cautiously about their financial habits. Each challenge is released by mentors, although the use of the app is encouraged by mentors and students alike through the positive reinforcement mechanisms. Also exciting is every mentor's ability to keep track of their individual students' progress through the mobile platform.*

The challenges allow students to accrue points and compete on a school-wide scale, permitting students from the same schools to see how many points their peers have accumulated. Students and mentors can scroll through their feeds and see what their peers and students are working on; they are even able to comment and like other peoples' posts. By "gamifying" the experience, students are more engaged and are more likely to participate.*

Challenges vary depending on the lessons that they accompany. An example of a challenge is the "SnapTrack"* challenge, in which students take pictures of things they spent money or saved money on. This challenge is meant to reinforce positive spending habits. Later in the curriculum, the "Good Buy, Bad Buy"* challenge becomes live and also reinforces positive spending. Students take pictures of the things they have purchased and other users vote on whether the item was a good buy or a bad buy. Thus, the challenges very much rely

* Financial Capability Curriculum, Moneythink.

on the interactions between users, and the app itself is an excellent way for mentors to keep in touch with their students throughout the week.

The following table, which the UChicago chapter sometimes distributes to potential school partners, contains a detailed list of Moneythink Mobile challenges. The mobile application utilizes hashtags to organize posts and make it easier for mentors and students to see what others are posting. Each challenge remains live for a week.

CHALLENGE NUMBER	CHALLENGE NAME	DESCRIPTION
1	Snaptrack	Snap photos while you're #spending and #saving this week. And be honest!
2	MiniGoal	Set a small financial goal (#minigoal) and celebrate when you accomplish it by posting to the feed.
3	The "In" Crowd	Use LinkedIn to make connections and show off your skills.
4	Savvy Selfie	Add a professional twist to your personal style and adapt your look for the working world.
5	Set to Save (1 of 3)	A key to financial success is making a budget. And sticking to it.
6	Staptrack to Save (2 of 3)	Think like a master money manager—a budgeter!— by phototracking this week using the spending categories you set for yourself.
7	Reality Check Your Budget (3 of 3)	Did you keep your spending in check last week? It's time to find out!
8	Windowshopping for a Bank	Checking? Savings? Big bank or check-cashing shop? Tap the wisdom of the crowd as you research to find the best option for you.
9	Brainstorm the Next Challenge!	Had fun with our challenges? Think you can do one better? Here's your chance. Post ideas for new challenges that teach financial literacy and how to save money. Now you're the expert #humblebrag.

Figure 3.3 shows what a challenge would look like once it has gone live and, more specifically, is an example of what a completed challenge looks like. Mobile allows for students and mentors to interact with one another. In this example, which is called the "Good Buy, Bad Buy" challenge, students take pictures of what they purchased and others vote on whether the purchase was a good decision.

The mobile application itself was developed through a partnership with IDEO.org, and a number of considerations and extensive

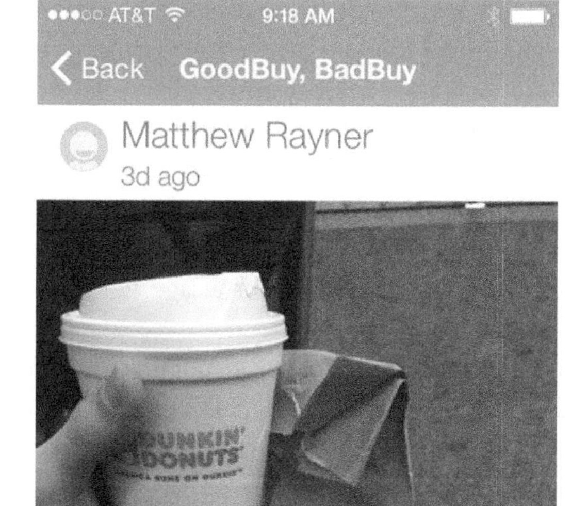

Figure 3.3 Good buy, bad buy challenge. One of the nine mobile challenges students participate in through the Financial Capability Curriculum.

research went into creating the actual phone application. More can be learned about the application's development through the Moneythink or IDEO.org web pages. Moneythink UChicago piloted the program in the spring of 2014, and it was incorporated into all other chapters beginning in the fall of 2014.

Metrics: How We Measure Success and Effectiveness

The Moneythink curriculum incorporates a number of strategies that will hopefully result in dynamic and engaging programming. Moreover, with the advent of Moneythink Mobile, mentors are able to stay connected with their students outside of the classroom as well. However, Moneythink Mobile provides another important opportunity for the National team—information.

With the phone application, we will be able to collect data about the spending habits of our mentees and tailor our lessons and curriculum into something that is even more relevant and hard hitting. Mobile will also be a key tool in measuring habit changes. We will be able to use the data from application usage not only for research purposes but also for metric purposes.

Are the lessons that we teach effective enough to lead to changes in spending and saving among our students? Which of our classrooms show the biggest changes? Questions such as these will be able to be answered from the data. Moreover, better and more effective mentoring strategies can be teased out by examining the mentors in the most effective classrooms. Moneythink Mobile provides the National team with a more unique method of measuring success without compromising quality.

The current method of determining effectiveness is fairly traditional. Pre- and postcurriculum tests are administered to see if and where changes in understanding have evolved. Furthermore, before the Mobile application, exit surveys were used to gauge understanding of material after each lesson.

We measure impact in three key categories. The first is knowledge gained—simply looking at the differences in understanding before and after our curriculum. The second vertical of impact is categorized as tangible output. Students are asked to create a number of things through the Moneythink program such as a LinkedIn profile and, most notably, a resume. This vertical is meant to measure what percentage of students completes our program with tangibles. The final vertical measures attitude change—the extent to which students are willing and able to change the way they approach money and personal finance after Moneythink.

This methodology, although generally successful, is not without its flaws. Most significantly, administering tests to understand changes in knowledge are most successful if the precurriculum and postcurriculums tests are administered at the true beginning and true ending of the Moneythink program. Unfortunately, this scenario is not always the case, as snow days, limited classroom time, and attendance consistency can be factors that cause tests to be administered later from the true start and end dates of the curriculum. In fact, more than anything, I believe that the impact of Moneythink in the scope of knowledge acquisition is understated given the number of times that I personally have seen the presurvey administered after the mentors have been in the classroom for some time.

What these tests also fail to measure are the specific changes in habits and decision making after curriculum implementation. Before mobile, we had no way of really knowing how our mentees changed the way they perceive and use money. We had our students' word of mouth and their individual stories, but no solid data to support the findings we had been told. This is why Moneythink Mobile is so revolutionary—it is completely changing the way in which Moneythink will measure success.

However, the statistical data that we do have are positive to say the least. A snapshot from the 2013–2014 annual report is shown in Figure 3.4, highlighting some of the key accomplishments from all of our chapters around the country.

Moreover, after Moneythink (Our Impact, Moneythink), the following changes were noted:

- Seventy-six percent of students believe they have the opportunity to become financially successful.

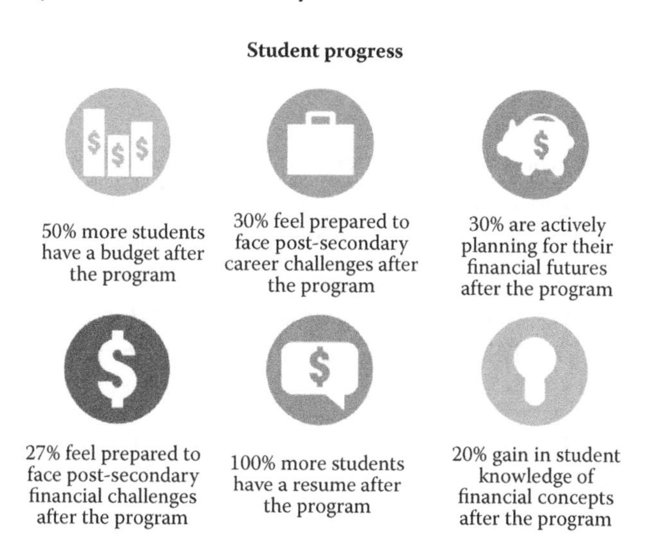
Student progress
50% more students have a budget after the program
30% feel prepared to face post-secondary career challenges after the program
30% are actively planning for their financial futures after the program
27% feel prepared to face post-secondary financial challenges after the program
100% more students have a resume after the program
20% gain in student knowledge of financial concepts after the program

Figure 3.4 Snapshot of Moneythink's success.

- A 49% increase in students' ability to manage their own finances.
- A 39% increase in students' belief that they will be more financially successful than their parents.
- A 48% increase in household discussions of money management.
- A 19% increase in students' confidence in their ability to make money.
- Eighty-six percent of students plan to graduate from college.

Future: Why Banking Is the New Initiative from the National Team

As the mobile application becomes more frequently used and better integrated into classrooms, the question remains regarding the future of Moneythink. What is the new initiative for the organization, and what is its goal for the coming year? It is clear that the new push from the National team is toward banking and encouraging our students to become part of the traditional banking system.*

So, what characterizes nontraditional forms of banking? Simply put, alternatives to the banking system fall into this category. Use of prepaid cards, payday loans, and check cashers are all examples of nontraditional forms of banking. And surprisingly these forms of banking are fairly popular, especially among Millennials (de Bassa Scheresberg et al., 2014).†

Figure 3.5 demonstrates the use of key products among the Millennial age group (Think Finance, 2012).

Shockingly, there is little difference between usage of these products among the lower and higher income groups. Regardless of income group, today's youth are likely to pursue nontraditional forms of banking.

Alternative financial services carry a lot of risks and tend to result in numerous fees, making the services costly. Therefore, it is astonishing that 34% of Millennials have used pawnshops or payday loans in

* This initiative was discussed at the 2014 Summer Leadership Institute held by the National team for all Moneythink campus leaders. This conference is the kickoff event for each new academic year where leaders from across different chapters come together to be retrained by the National team as well as collaborate in regards to effective mentoring techniques.

† Millennials range in age from 23 to 35 years.

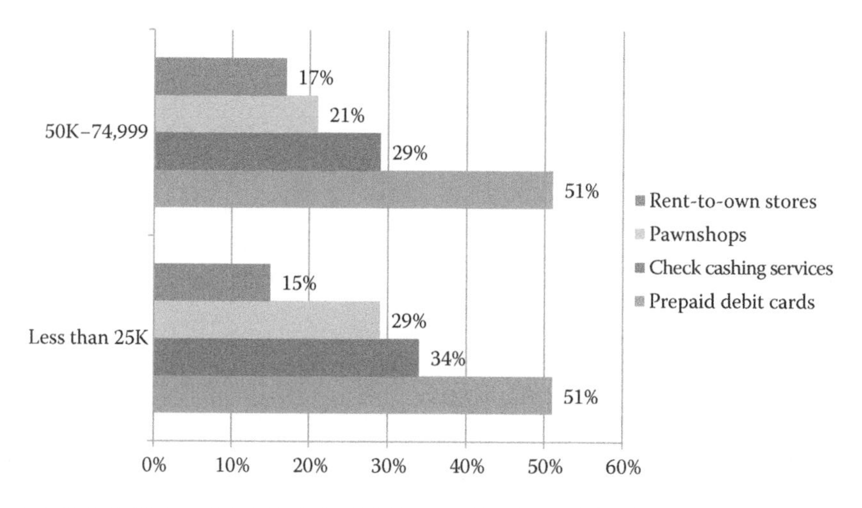

Figure 3.5 This chart compares the use of specific nontraditional banking products among Millennials among income groups. There is not much difference with products used, regardless of income group.

the past 5 years, and of this subgroup, 52% of those who have used these services have done so three or more times in the past 5 years (de Bassa Scheresberg and Lusardi, 2014).

The jury is out—younger generations are using risky alternative financial services and as an organization tailored to the youth, Moneythink is doing what it can to prevent our students from pursuing these options (Allgood and Walstad, 2013).

However, race, education levels, and income levels also affect the use of nontraditional banking products, and the long-term effects could be dangerous. Statistics from the state of Illinois (the state where Moneythink UChicago specifically operates) are included in the following paragraphs. Hopefully, they will provide numerical insight for the importance of encouraging use of traditional banking systems.*

Initially, the data regarding banked versus unbanked within Illinois (Figure 3.6) do not seem problematic. The vast majority of residents (more than 70%) are fully banked.

* For a more human approach to the banking system, consider spending about 40 minutes watching "Spent: Looking for Change," a documentary available online created by the Young Turks online news organization. It explores the stories of individuals who are lost to the banking system and very effectively explores the issues with alternative financial services and how families can become stuck in the alternate system.

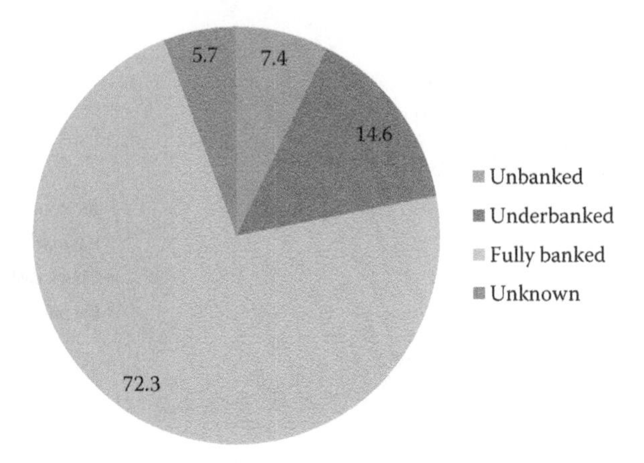

Figure 3.6 Overall banking statistics for Illinois residents (in percent). The majority of Illinois residents are fully banked. (From Burhouse, S. et al., 2013 FDIC National Survey of Unbanked and Underbanked Households, 2014. With permission.)

However, when we start to look at data by race, the issue becomes much more problematic (Figures 3.7 through 3.10).

The disparity is stark: minorities have much lower percentages of being fully banked compared to the White non-Black, non-Hispanic population. In fact, for African Americans and Hispanics, the unbanked and underbanked population hovers about 46%, whereas for the Caucasian population the unbanked and underbanked subgroup is about 13% in the state of Illinois (nearly 3.5 times smaller compared to African Americans and Hispanics; Burhouse et al., 2014).

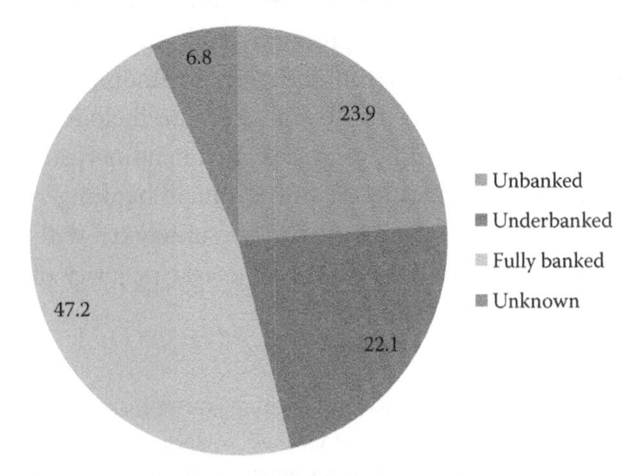

Figure 3.7 The Black population of Illinois' banking statistics (in percent). Only 47% of the Black population of Illinois is fully banked. (From Burhouse, S. et al., 2013 FDIC National Survey of Unbanked and Underbanked Households, 2014. With permission.)

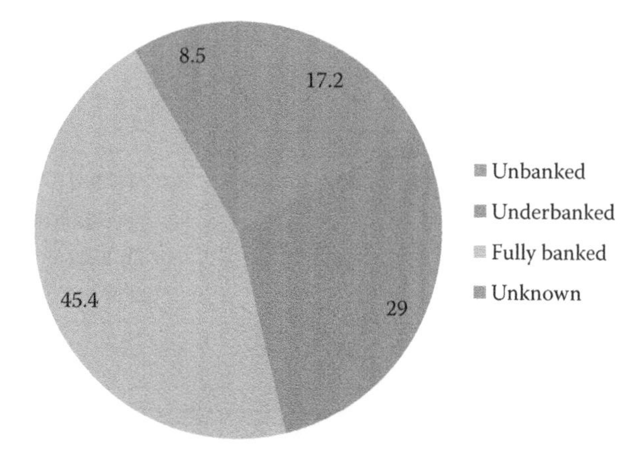

Figure 3.8 The Hispanic population of Illinois' banking statistics (in percent). Only 45% of the Hispanic population of Illinois is fully banked. (From Burhouse, S. et al., 2013 FDIC National Survey of Unbanked and Underbanked Households, 2014. With permission.)

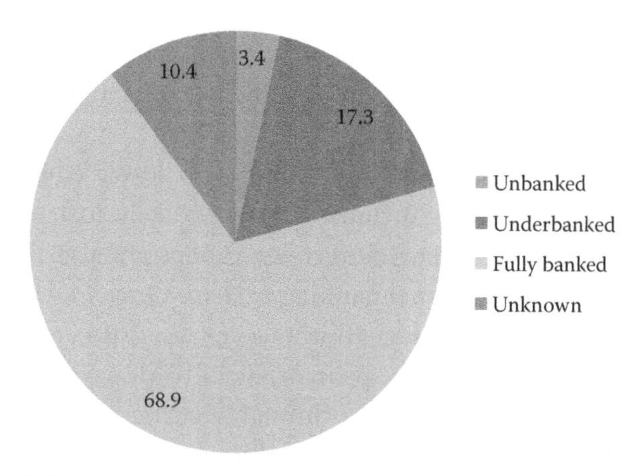

Figure 3.9 The Asian population of Illinois' banking statistics (in percent). Nearly 69% of the Asian population of Illinois is fully banked. (From Burhouse, S. et al., 2013 FDIC National Survey of Unbanked and Underbanked Households, 2014. With permission.)

The blatant differences continue when exploring education and income levels. Nearly 54% (53.9% to be exact) of the "less education" population—those without a high school diploma—are unbanked or underbanked. This number drops precipitously to 9.8% of unbanked and underbanked among the population with a college degree, with only 0.8% of this subgroup being unbanked. Families that make less than $15,000 a year have an unbanked and underbanked population of 47.4% compared to families that make at least $75,000 a year with

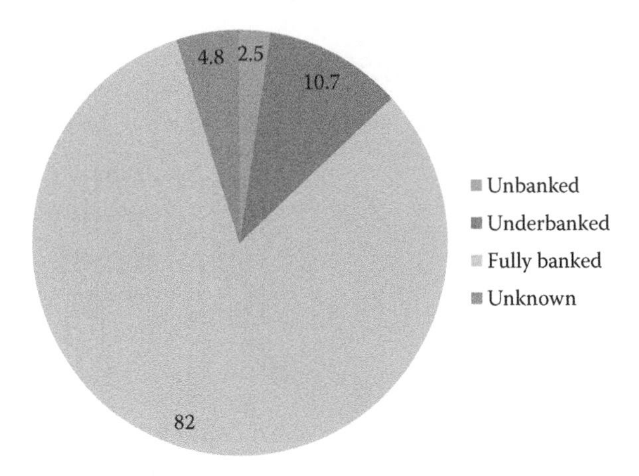

Figure 3.10 The White (non-Black, non-Hispanic) population of Illinois' banking statistics (in percent). Eighty-two percent of the White population of Illinois is fully banked. (From Burhouse, S. et al., 2013 FDIC National Survey of Unbanked and Underbanked Households, 2014. With permission.)

an unbanked and underbanked population of 9.4% (Burhouse et al., 2014). These statistical figures are specific to the state of Illinois but are mirrored across the nation.

The trend remains: minorities, those with lower education levels, and those with lower incomes are more likely to be unbanked or underbanked. Moreover, younger age groups are also likely to be excluded from the traditional banking system. Only 44.4% of 15- to 24-year-olds are fully banked (Burhouse et al., 2014). The need to address this issue is painfully apparent; luckily, Moneythink is working hard to address this situation through a curriculum focused very heavily on the merits of traditional banking systems, the importance of strong credit, and the high costs associated with untraditional banking methods.

Moneythink Is: What Being a Mentor Means to Me

When I reflect on my time at the UChicago, Moneythink is always the first thing that I think about. The need for financial literacy education is obvious. Only six out of 50 states require schools to test financial literacy, and thus, really only six states are enforcing personal finance knowledge in U.S. high schools (Economic and Personal Finance Education in Our Nation's Schools 2014, Council for Economic Education). And yet, managing money and learning about the ways financial institutions work are perhaps the simplest way to provide economic opportunities for students.

But it is more than the need for these conversations about money that make my experiences with Moneythink so memorable. The students I have taught have been nothing but incredible. Eager to learn and ask questions, my mentees have forced me to expand my knowledge base through their inquisitiveness. The administrators I have worked with have been nothing short of unbelievable—dedicated to their students and the importance of bringing financial knowledge to them. Most importantly, the relationships you are able to build with your mentees are absolutely unparalleled. I have had coffee with my students outside of the classroom and have given them tours around my own university campus.

Our chapter has been able to engage with our students and provide additional programming such as speakers and entrepreneurship competitions. This year, we brought students from all over the city of Chicago to participate in "Money Tank," our chapter's version of Shark Tank and a platform for our students to participate in an entrepreneurship challenge and meet other students from the city. Money Tank is an opportunity that is unique to our chapter. However, the National team has been nothing but supportive in making this dream of our chapter come to fruition. The creativity and control over programming that the National team allows makes mentoring with Moneythink even more rewarding. We have control of what we teach and the way the lessons we create are taught. We are able to address the needs of our students and are not restricted by a dominating national organization. This has allowed mentoring with Moneythink to be so much more than just financial literacy education—it is relationship building.

In addition, the other people at my own university that I have been lucky enough to meet through mentoring are absolutely incredible. I am convinced that the university students involved with the Moneythink Movement are some of the best and brightest in the world. Incredibly passionate, intelligent, relatable, and fun loving, Moneythink mentors are universally some of the best people I have met during my undergraduate years. The people that inspire me most on the UChicago campus are people I have met through Moneythink. Our UChicago chapter has investment bankers, chemistry majors, and a Truman Scholar. We have White House interns, fraternity men, and athletes. But most importantly, we have a group of incredibly dedicated and diverse mentors who never give less than 110% in the classroom.

When I reflect back on my college career, Moneythink has undoubtedly been the most important and life-changing experience I have had. I am proud to be a Moneythinker today and a Moneythinker for life. We are Moneythink, and we are already changing the world of financial literacy education.

References

Allgood, S., and Walstad, W. (2013). Financial Literacy and Credit Card Behaviors: A Cross-Sectional Analysis by Age. *Numeracy, 6*(2).

Burhouse, S., Chu, K., Goodstein, R., Northwood, J., Osaki, Y., and Sharma, D. (2014). 2013 FDIC National Survey of Unbanked and Underbanked Households.

Carpenter, J. (2014, March 7). What a Chicago Financial Nonprofit Is Doing to Make Itself Become Obsolete. Blue Sky Innovation. *The Chicago Tribune*. Retrieved from http://bluesky.chicagotribune.com/originals/chi-ted-gonder-moneythink-app-bsi-20140305,0,0.story.

Council for Economic Education. (n.d.). Survey of the States: Economic and Personal Finance Education in Our Nation's States.

de Bassa Scheresberg, C. (2013). Financial Literacy and Financial Behavior among Young Adults: Evidence and Implications. *Numeracy, 6*(2).

de Bassa Scheresberg, C., and Lusardi, A. (2014). Gen Y Personal Finances: A Crisis of Confidence and Capability.

de Bassa Scheresberg, C., Lusardi, A., and Yakoboski, P. J. (2014). College-Educated Millennials: An Overview of Their Personal Finances.

History. (n.d.). Retrieved from Moneythink: http://moneythink.org/about-us/history/.

Lusardi, A., and Mitchell, O. S. (2011, June). Financial Literacy and Retirement Planning in the United States. *National Bureau of Economics Working Paper Series*.

Mint. (2015, April). Expert Interview with Ted Gonder of Moneythink on Financial Education for Young People. Retrieved from https://www.mint.com/expert-interview-ted-gonder-moneythink-financial-education-young-people#sthash.xEWYpjz7.dpuf.

Moneythink. (n.d.). Our Approach. Retrieved from http://moneythink.org/our-program/our-approach/.

Moneythink. (n.d.). Our Impact. Retrieved from http://moneythink.org/our-program/impact-numbers/.

Moneythink. (n.d.). Our Program. Retrieved from http://moneythink.org/our-program/.

Think Finance. (2012, May 17). Millennials Use Alternative Financial Services Regardless of Their Income Level.

Woodyard, A., and Robb, C. (2012). Financial Knowledge and the Gender Gap. *The Journal of Financial Therapy, 3*(1).

4

The Financial Education Movement in Pennsylvania

Fifteen Years and Still Going

HILARY HUNT AND
CATHY FAULCON BOWEN

Contents

Introduction

Financial education efforts and outcomes vary widely from state to state. Recurring surveys and reports including the Survey of the States from the Council on Economic Education (2014) bear this out with lists of states that include personal finance content in their states' academic standards and/or require a personal finance course for graduation from high school. These reports provide a snapshot of financial education at a given point in time. However, to paint a full picture of the status of financial education in any given state at one time, more information is needed. This is particularly true in states such as Pennsylvania that have high levels of local control over education.

Making progress to improve youth financial literacy in a state that highly values local control of education is a tall order and one that has been ongoing in Pennsylvania for more than 15 years. During this time, a series of diverse initiatives has seen varying levels of success. In this chapter, we will explore the array of efforts, the impact they have made on financial education, and some lessons that have been learned.

Framing Financial Literacy in State Education Policy

States enact and oversee education policy in many ways. Such policies impact every level of education—including efforts to improve financial literacy. Thus, to understand the status of financial education in Pennsylvania, it is helpful to first frame it in the context of state education policy. Although all states have some version of statewide academic standards and testing in core subject areas, the subjects and grade levels vary from state to state. Pennsylvania was the 48th state to adopt statewide standards with the implementation of Chapter 4 (Academic Standards and Assessment) in 1999. Although the state does not have stand-alone content standards for personal finance, there is considerable coverage of such topics within other content areas. The first such standards appeared in the state's academic standards for economics and for family and consumer sciences, which took effect in 2003. In 2004, additional personal finance content was required with the adoption of standards for Career Education and Work. These were bolstered by the implementation of voluntary standards for Business, Computer, and Information Technology (BCIT) in 2012.

Although there is significant personal finance content in the state standards, implementation and assessment of these standards are left to each of Pennsylvania's 500 school districts. There is little oversight from the state level. The same holds for graduation requirements. Each of Pennsylvania's 500 locally controlled school districts sets its own graduation requirements, leading to considerable variation among districts not only with regard to personal finance but other related subjects including social studies/economics, family and consumer sciences, and business. Without a set of statewide standards for high school graduation, there has been no requirement in Pennsylvania for every high school student to take a course in personal finance for graduation.

There are districts, though, that recognize the need for financial education and require a stand-alone course in personal finance for graduation. Determining which of the 500 school districts have this graduation requirement is challenging. School districts are not required to report this information to the Pennsylvania Department of Education (PDE), and gathering it from each district is laborious. The first such effort was made in 2007 through a cooperative effort of the state's Office of Financial Education (Office) and the Pennsylvania State University's Survey Research Center and repeated again in 2009 (Penn State University Survey Research Center, 2008, 2010). A 2013 report to the legislature from PDE gathered similar information (Pennsylvania Department of Education and Pennsylvania Department of Banking and Securities, 2013). Although Table 4.1 lists the number of school districts requiring a stand-alone course in personal finance has increased, it is still far from comprehensive. Furthermore, Table 4.2 shows that school districts requiring a course in personal finance in 2013 were concentrated in 18 of Pennsylvania's 67 counties. With more than 450 school districts still lacking such a graduation requirement, it is unlikely that this many will voluntarily make personal finance a graduation requirement without a state mandate. Thus, state legislative or regulatory change will be needed for Pennsylvania's high school graduates to be exposed to personal finance. Although there is widespread agreement of the need for financial education, a number of roadblocks have prevented this from occurring.

Opponents of such a mandate cite a number of concerns—most having little to do with the need for increased financial education. First, requiring a course in personal finance for graduation would be the first attempt by the state to mandate coursework beyond that required for assessment. Supporters of local control consider this a "slippery slope"

Table 4.1 School Districts in Pennsylvania Requiring a Stand-Alone Course in Personal Finance for Graduation

YEAR	2007	2009	2013
Number of school districts with graduation requirement	20	44	38

Table 4.2 Pennsylvania Counties with One or More School Districts with a Stand-Alone Course in Personal Finance for Graduation in 2013

COUNTY	NUMBER OF SCHOOL DISTRICTS WITH A PERSONAL FINANCE COURSE
Adams	2
Allegheny	1
Armstrong	1
Chester	1
Columbia	3
Cumberland	2
Delaware	1
Erie	1
Fayette	1
Franklin	1
Fulton	1
Huntingdon	1
Lancaster	4
Montgomery	2
Schuylkill	2
Snyder	1
Wayne	1
Westmoreland	1

Source: Pennsylvania Department of Education & Pennsylvania Department of Banking and Securities, 2013. Economic and Personal Finance Education in Pennsylvania: A Report to the Governor and General Assembly. Retrieved from https://www.portal.state.pa.us/portal/server.pt/document/1332884/financial_education_report_pdf.

toward requiring other coursework that some districts may not find necessary. Others are concerned with the resources needed to make such a requirement a reality including having enough well-qualified teachers to instruct courses and having sufficient curricular resources for quality instruction. The latter has been of increased concern since 2011, when significant reductions to the state's education budget were made. Between the 2010–2011 school year and the 2014–2015 school year, these reductions led to the elimination and reduction of more than 1,150 academic programs and the loss of more than 23,000 education positions (Buckheit and Himes, 2014). Among these were the elimination of entire business and family consumer sciences programs (which tend to be the areas where personal finance courses are taught) and the associated teacher positions. The state's most recent report on the status of economic and personal finance education indicated that in the 8-month period from April 2011 to November 2011, 22 school districts requested to curtail or eliminate their family and consumer sciences program whereas 24 made similar requests for their business programs (Pennsylvania Department of Education & Pennsylvania Department of Banking and Securities, 2013).

Understanding the constraints of the educational policy landscape and the effect it brings to efforts to increase financial literacy endeavors is vital in understanding efforts to improve youth financial literacy in Pennsylvania and elsewhere. Without understanding the influence of educational policies, efforts to improve financial literacy may seem illogical or misdirected. This is especially true when comparing efforts in states with significant statewide control over education such as Texas or Florida to states such as Pennsylvania, wherein such efforts are more locally controlled. Adding a course in personal finance to an existing list of the state graduation requirements is a very different task than having financial literacy made the only requirement beyond those required for assessment.

Setting a Course for Action: Formation of the Pennsylvania Jump$tart Coalition

The first coordinated effort to improve youth financial literacy in Pennsylvania occurred in 1999 when the Pennsylvania Jump$tart Coalition for Personal Financial Literacy was formed. The Pennsylvania Coalition was one of the first state coalitions to be formed by the national Jump$tart organization. In part, state coalitions were formed because leaders recognized that education policy decisions—including those affecting financial literacy—are made at state and local levels. Thus, state coalitions were formed to help organize and coordinate efforts to improve financial literacy and, when possible, to further efforts to bring about legislative and policy changes. Although coalitions in other states attempted to influence such change through legislative efforts, Pennsylvania's Coalition recognized the locally controlled education policy climate and took an alternate course.

The Pennsylvania Jump$tart Coalition convened representatives from organizations interested in financial literacy and developed a common vision for financial education. The Coalition determined its initial focus would be to support the inclusion of personal finance in academic standards proposed by the state and to encourage coordinated efforts to prepare teachers to provide quality financial literacy instruction. To this end, Coalition members testified at public hearings on various state standards and coordinated efforts to provide professional development for teachers in conjunction with established teacher conferences. The Coalition also raised awareness of the need for financial education through public awareness efforts including requesting that April be proclaimed Financial Education Month. Financial Education Month was proclaimed for the first time in 2000, and the designation has been reaffirmed every year since by five governors.

In 2004, the Pennsylvania Jump$tart Coalition was the first organization in the country to receive a grant from the Investor Protection Fund. This fund was administered by the Investor Protection Trust and was established because of a settlement of enforcement actions involving conflicts of interest between research investment banking operations. Under this grant, the Coalition prepared teachers across Pennsylvania during a series of *Money Counts Conferences.* More than 500 teachers participated in these 1-day sessions that featured speakers on topics ranging from trends in securities fraud and abuse to innovative ways to teach students about investing. The conferences brought together many Jump$tart partners to offer the trainings and featured the curriculum resources of various program providers.

A Model for Other States: Pennsylvania's Office of Financial Education

Efforts of the Coalition were significantly bolstered in 2004 when Governor Edward G. Rendell established by the Governor's Task Force for Working Families the Pennsylvania Office of Financial Education via an executive order. This Task Force conducted 24 public roundtables across the state to gather information on areas important to working families. Four key areas emerged from the statewide roundtables and were offered as recommendations for efforts to support working families: improve financial literacy, build assets, prevent financial abuse, and increase incomes for working families. The Task Force's recommendations helped establish a plan for the Office to achieve its mission of improving the quality and quantity of financial education across the state.

With the governor's imprimatur to work with and for all state agencies, the Office was uniquely positioned to influence financial education in the state. Housed within the state's Department of Banking, the Office was funded through the regulatory fees and fines placed on the state's financial institutions rather than through the general fund of the state. Although this allowed the Office to grow from a staff of one to five, its resources were still limited and would not allow a "direct-to-consumer" approach. Instead, the Office established a three-pronged approach to financial education focused on delivery systems: schools, workplaces, and community-based organizations with a staff member dedicated to each area. It was the most comprehensive and largest effort to date by any state government to address financial education.

The Office's school-based financial education efforts were unique with a strong partnership between the Departments of Banking and Education. This partnership included a shared staff member and memorandums of understanding that allowed coordinated funding efforts. Much of the school-based efforts focused on teacher professional development. The Office supported the integration of personal finance topics in elementary and middle school coursework while advocating for stand-alone courses at the high school level. In its 7 years of operation, the Office worked with more than 2500 teachers and administrators from across the state to improve financial literacy. The Office is credited with much of the increase in schools offering courses in personal finance and many even requiring such a course for graduation. Helping teachers and administrators make the case for financial literacy courses, providing training and materials at no charge to districts, and connecting schools with high-quality and free materials were key in these efforts.

The hallmark of the Office's school-based program was its Governor's Institute for Financial Education. Conducted as a partnership between the Departments of Education and Banking, this 1-week, residential program provided curriculum resources and improved content knowledge for more than 500 teachers each summer from 2005 to 2011. Throughout the week, teachers learned ways to incorporate personal finance into their instruction and develop stand-alone courses from state and national experts including many members of the Jump$tart Coalition. Partnerships with other agencies allowed the teachers to improve their own personal finances with a session on credit reports and opportunities to discuss their personal finances with financial experts. In addition, the Institute served as an opportunity to seed local policy changes for financial literacy. After attending the Institute, many teachers made it their mission to increase student access to personal finance courses in the districts where they taught and/or lived.

In addition to the school-based efforts, the Office formed partnerships with other organizations to advance financial education. The Office assisted with the development of regional coalitions in Pittsburgh and Philadelphia, coordinated by the Federal Reserve Banks of Cleveland and Philadelphia, respectively. These local efforts brought together community-based organizations in each region for several years to exchange ideas and coordinate efforts. As with many financial education collaboratives, these groups flourished for a short time but lacked staying power. Like similar financial literacy efforts, local collaboratives were unsuccessful, at least in part, owing to a lack of focus and lack of dedicated leadership whose main purpose was to maintain the partnerships. Without a strong focus on the key issues, financial literacy initiatives can quickly expand to secondary matters and be easily derailed.

A partnership with the Pennsylvania State University's Cooperative Extension was one of the Office's most successful efforts to improve financial literacy for youth through community-based organizations. Although many materials existed for schools, there were few programs that sought to provide financial education for families. A program by the local public broadcasting station in Erie, Pennsylvania, served as a model for the development of *Right on the Money: Talking Dollars and Sense with Parents and Kids*. Funded through a grant from the Heinz Endowments, the *Right on the Money* program was developed and piloted in partnership with libraries across Pennsylvania to reach parents with children ages 5–7. Since its launch in 2008, the program has been conducted in numerous libraries and schools throughout the state and has been adopted in other states. It has also been translated into Spanish for use with additional audiences. In 2009, the curriculum received the Excellence in Financial Literacy Education Award for Post-Secondary/ Adult Curriculum of the Year from the Institute for Financial Literacy.

For its efforts to improve financial literacy through schools, workplaces, and communities, the Office saw great success and recognition during its years in operation. The Office was considered a model for statewide financial literacy efforts. Representatives of the Office were frequently invited to participate in national summits and other events focused on improving financial education across the country including those hosted by the U.S. Treasury, the White House, the U.S. Government Accountability Office, and Consumer Financial Protection Bureau. In 2010, its efforts were honored with a "Bright Idea Recognition" from Harvard University's Ash Center for Democratic Governance and Innovation in the John F. Kennedy's School of Government, and an Excellence in Financial Literacy Education Award for Organization of the Year from the Institute for Financial Literacy. In 2011, however, following the inauguration of a different governor, new priorities led to the Office's closure in November of that year and the elimination of the Director of Financial Education position in 2013.

Legislative Efforts to Improve Financial Literacy

Before 2010, all governmental efforts in Pennsylvania to improve financial literacy were led by the executive branch through regulatory changes (such as the inclusion of personal finance in the state's academic standards) or through programmatic efforts such as those led by the state's Office of Financial Education. Act 104 of 2010 was the first legislation in Pennsylvania to address the need for financial education. Met with bipartisan support, the act required the PDE, among other things, to accomplish the following: (1) provide curriculum materials and financial literacy resource information to educators and public and private schools and organizations; (2) convene a task force to assess trends and needs in financial literacy, consider the manner in which funds are used to support financial education, and make recommendations for ways to improve financial education; and (3) issue a biennial report on the status of financial literacy programs in the commonwealth highlighting new initiatives and recommending future program needs (Public School Code of 1949—Omnibus Amendments, 2010).

Pursuant to this legislation, in 2011 the PDE established the Task Force on Economic Education and Personal Financial Literacy Education and appointed its nine members as outlined in the act. Members represented the education sector with a teacher, school superintendent, school board member, and university faculty, and the financial services sector with representatives from the banking, credit union, and mortgage industry along with the Federal Reserve Bank of Philadelphia and the Pennsylvania Department of Banking and Securities. After 14 months of research and meetings, the Task Force issued a report with key findings and recommendations (Pennsylvania Task Force on Economic Education and Personal Finance Literacy Education, 2013). The recommendations urged the Governor and General Assembly to

1. Require every Pennsylvania high school student to complete a stand-alone capstone course on personal finance in order to graduate
2. Adopt comprehensive, stand-alone Pennsylvania K–12 academic standards devoted to personal finance
3. Provide dedicated funding to support high-quality K–12 personal finance instruction and teacher education
4. Develop a financial education instructional endorsement for secondary teachers in Pennsylvania and corresponding program guidelines for professional educator programs

Since the Task Force report was issued in January 2013, additional legislation has been introduced to address some of these recommendations. Two bills were introduced in the 2013–2014 legislative session including one that would have made a course in personal finance a requirement for graduation. In October 2014, the Education Committee of the Pennsylvania House of Representatives held a public hearing on financial literacy that featured testimony from 13 individuals and organizations, each voicing tremendous support for financial education. As of April 2015, three bills had been introduced for the 2015–2016 session between the House and the Senate. Although these efforts have significant bipartisan support, each faces significant opposition as previously described because of local control and funding to support implementation.

The Making Cents Project

In addition to the formation of the Task Force on Economic Education and Personal Financial Literacy Education, the PDE embarked on a partnership with the Pennsylvania State University in January 2012 to meet the other requirements of Act 104 related to financial literacy. The resulting endeavor came to be known as The Making Cents Project. This project has provided curriculum resources and teacher professional development for schools and organizations across Pennsylvania. Each school year since, The Making Cents Project has hosted a series of webinars featuring information on personal finance topics, ready-to-use classroom tools and resources, and professional information for teachers of personal finance. The webinars have provided support to more than 300 Pennsylvania teachers with the majority teaching at the secondary level. Participation in the live webinars provides teachers with continuing education credits, and past webinars can be accessed and viewed online via the project's website: http://www.makingcentspa.org.

To support the teaching of personal finance in classrooms at all grade levels, a PreKindergarten–Grade 12 (PK–12) Personal Finance Instructional Framework was made available in the fall of 2014 along with a Model High School Personal Finance course. The PK–12 Instructional Framework identifies long-term transfer goals along with six big ideas and associated essential questions for financial education across the grade levels. At each grade band (PK-2, 3–5, 6–8, and 9–12), key concepts and competencies are identified and correlated to state standards. These are matched with suggested strategies to use in integrating personal finance concepts into mathematics, English language arts, social studies, and general classroom management in grades PK–8. The Model High School Personal Finance course builds on the instructional framework providing outlines for six modules with sufficient lessons and instructional strategies to fill a one-semester stand-alone course in personal finance. Both documents are available for educators to access through PDE's Standards Aligned System portal at http://www.pdesas.org.

Other project components include the distribution of information through an electronic newsletter and the development of additional resources to support instruction. An iTunes U course in personal finance will be made available through PDE's *Pennsylvania Learns* partnership with iTunes during the 2015–2016 school year. Future efforts are expected to focus on further integration of mathematics and personal finance concepts both in instruction and assessment and the development of an endorsement in personal finance for previously certified teachers. These efforts are in keeping with the recommendations PDE made in conjunction with the Department of Banking and Securities in the 2013 report on the status of personal finance and economic education and support some of the recommendations of the Task Force.

Who's Who in State Financial Education Efforts

Besides the state's education and financial regulatory agencies, others in state government and beyond serve important roles in supporting financial education. In Pennsylvania, these include the Pennsylvania Higher Education Assistance Agency, which serves as the guarantor and servicer of student loans and conducts programs for schools, parents, and students in order to support completion of the free application for federal student aid. The Pennsylvania Securities Commission has a long-standing commitment to investor education that overlaps with financial education. Through in-person presentations to students and the dissemination of curriculum resources, the commission works to promote safe saving and investing behaviors. The commission merged with the Department of Banking in 2013, and the merged department continues many of these efforts. Other agencies that have been involved over the years include the Insurance Commission, the Pennsylvania Housing Finance Agency, the Department of Military and Veterans Affairs, the Treasury Department, and the Office of Attorney General.

Outside of state government, there is significant involvement in financial literacy from the private and nonprofit sectors. The Pennsylvania Credit Union Association maintains a separate foundation, which is instrumental in supporting financial education efforts in schools. The association's efforts include making grants to member credit unions to support the development of branches in high schools and supporting meetings where those with interest in financial education coalesce. The Pennsylvania Bankers Association promotes Teach Children to Save Day each April to encourage bankers to volunteer and teach personal finance lessons in schools. Individual banks and credit unions have also developed signature programs for consumers in their target markets.

Nonprofit organizations such as Junior Achievement, Economics Pennsylvania, and Operation Hope also offer financial education programs on a regional basis throughout the state. Although each of those organizations has considerable reach and program penetration in some areas of the state, none are comprehensive statewide. In recent years, the United Way has taken a new leadership role in promoting financial literacy through community-based organizations.

Equipping Teachers with Knowledge and Tools

Many of the aforementioned programs including the Money Counts Conference, the Governor's Institute on Financial Education, and the Making Cents Project include professional development for teachers. Professional development for teachers is also a fundamental offering of the Federal Reserve Bank of Philadelphia, which offers a range of programs for teachers from evening seminars to weeklong summer programs. All of these programs have similar aims: improving both the content knowledge and the pedagogical skills of teachers as it relates to personal finance.

Among the challenges that financial literacy movements encounter are the lack of well-trained and informed teachers who are able to implement quality programs in schools. Both national surveys of teachers such as those conducted by the National Endowment for Financial Education (Holden and Way, 2010) and one of Pennsylvania's teachers conducted by Pennsylvania State University in 2008 on behalf of the Pennsylvania Securities Commission and with funding from the Investor Protection Trust (Hunt, 2009), indicate that teachers lack fundamental knowledge and skills needed to teach personal finance. It is for this reason that many of the efforts in Pennsylvania have included professional development as a component. Providing teachers with the knowledge and skills they need to effectively teach personal finance is essential to ensuring that students receive quality financial education. Similarly, efforts to develop curriculum resources have sought to fill gaps in the availability of resources such as the Right on the Money program for families and the PK–12 Instructional Framework and Model High School Personal Finance course for schools.

Hurdles to Overcome

Although much work has been done to improve youth financial literacy in Pennsylvania over the past 15 years, much work remains to be done. In addition to the aforementioned local control and funding challenges, there are still others. Although there is no lack of supporters of financial education, there is a lack of consistent leadership, funding, and high-level commitment. Financial education efforts need to achieve staying power so that such efforts do not come and go with each successive administration or party change in either the executive or the legislative branch.

The competing interests of groups involved in financial education can also work against one another to the detriment of the cause as a whole. Such was the case in 2013 and 2014 when the association representing business education teachers lobbied the Pennsylvania House of Representatives to support legislation (HB 1739) that would have authorized only those with a business certification to teach personal finance, thus effectively cutting off those programs taught in family and consumer sciences, mathematics, or social studies departments. Competition between financial institutions can have similar negative impacts when school districts avoid making a choice between collaborating with one institution over another and simply forego a financial education program as a result.

To get ahead of some of these challenges will require proactive measures that may be difficult to accomplish. Obtaining consistent and reliable data is another challenge to financial education efforts. School districts are not required to report their policies or actions related to financial literacy, and requiring such reporting may be perceived as an undue regulatory burden. Fully integrating personal finance and mathematics—especially in assessment—may be feasible and could make a significant difference in furthering the cause of financial literacy, albeit at a substantial price tag to implement. Similarly, a long-term approach to addressing the knowledge and skill gaps of teachers is to make sure that teacher preparation programs adequately prepare teachers in all subject areas to integrate personal finance. This, too, would require substantial and systemic changes and carry the anticipated costs along with them.

Other challenges transcend the local and state level and are national in scope. The federal government has increased the prominence of financial literacy through the establishment of the Financial Literacy and Education Commission and the President's Advisory Council on Youth Financial Capability. However, the subject of financial education has yet to make it to the top of the agenda of educational policy leaders and elected officials to bring it on par with topics such as STEM programs (science, technology, engineering, and mathematics), environmental education, or drug and alcohol awareness programs. Furthermore, financial education has not received funding equal to any of these areas.

Additional challenges are also outside of the scope of government at any level. Although states and local policymakers have the ultimate say in graduation requirements, the power of the National Collegiate Athletic Association (NCAA) is not to be overlooked. Student athletes and the high schools that prepare them pay careful consideration to its clearinghouse and eligibility guidelines. Currently, these guidelines clearly stipulate that courses in personal finance or consumer economics may not be counted toward eligibility. This challenge, however, can be viewed just as much as an opportunity. If the NCAA were to reverse this policy and go a step further to require all students to take a course in personal finance for eligibility, schools would undoubtedly offer and require these courses.

Finally, measuring the impact and success of financial education mandates or the implementation of voluntary standards is a long-standing challenge. Ideally, a longitudinal study would compare recipients of financial education from well-qualified teachers to those receiving similar education from teachers with no specialized training or those in a control group. Furthermore, comparing the impact of specific programs and curriculum would be ideal. Such a study, though, would require massive amounts of resources and has yet to be conducted. Instead, efforts to prove the efficacy of financial education programs compare students well after the instruction is received and cannot determine the extent or quality of the programs or the instructors.

Moving the Ball Forward

In order for financial education to achieve wide-scale adoption and success in Pennsylvania, and arguably in other states, leaders must acknowledge both the need and the challenges and move forward with coordinated and cooperative efforts. Such efforts need to embrace the desire to achieve long-standing impact and set aside political motives. Attacking the issue from more than one front is also helpful. From a policy standpoint, this means pushing for change at the local level—no matter how slow and arduous that may be (one district at a time for 500 school districts is rather arduous)—while simultaneously working for state-level impact such as a statewide graduation requirement. Similarly, efforts to improve the quality of instruction in financial education should be addressed in programs preparing future teachers as well as ongoing professional development programs for existing teachers.

One thing is for certain. Each year as Pennsylvania decision makers (e.g., legislative branch, department of state government, or local school boards) with the legal authority to make decisions in the best interest of Pennsylvanians wrangle with balancing the costs and benefits of providing a comprehensive approach to financial education, the cost to our future wage earning investors for taking no actions increases. With each passing year, another generation of children in Pennsylvania moves into adulthood without a solid base of financial knowledge and skills to manage their personal finances.

References

Buckheit, J. and Himes, J. (2014). Continued Cuts: The Fourth Annual PASA–PASBO Report on School District Budgets. Pennsylvania Association of School Administrators & Pennsylvania Association of School Business Officials. Retrieved from http://www.pasa-net.org/budgetreport6-5-14.pdf.

Council on Economic Education. (2014). Survey of the States: Economic and Personal Finance Education in Our Nation's Schools 2014. Retrieved from http://www.councilforeconed.org/wp/wp-content/uploads/2014/02/2014-Survey-of-the-States.pdf.

Holden, K. and Way, W. L. (2010). Teachers' Background and Capacity to Teach Personal Finance: Results of a National Study. National Endowment for Financial Education. Retrieved from http://www.nefe.org/Portals/0/WhatWeProvide/PrimaryResearch/PDF/TNTSalon_FinalReport.pdf.

Hunt, H. (2009). *Pennsylvania Teacher Investor Education Research Project: Report Findings for the Pennsylvania Securities Commission.*

Penn State University Survey Research Center for the Pennsylvania Office of Financial Education. (2008). Status of Financial Education in Pennsylvania Schools 2007. Retrieved from http://www.moneysbestfriend.com/download.aspx?id=241.

Penn State University Survey Research Center for the Pennsylvania Office of Financial Education. (2010). Status of Financial Education in Pennsylvania Schools 2009. Retrieved from http://www.moneysbestfriend.com/download.aspx?id=556.

Pennsylvania Department of Education & Pennsylvania Department of Banking and Securities. (2013). Economic and Personal Finance Education in Pennsylvania: A Report to the Governor and General Assembly. Retrieved from https://www.portal.state.pa.us/portal/server.pt/document/1332884/financial_education_report_pdf.

Pennsylvania Task Force on Economic Education and Personal Finance Literacy Education. (2013). Report and Recommendations. Retrieved from http://www.portal.state.pa.us/portal/http;//www.portal.state.pa.us;80/portal/server.pt/gateway/PTARGS_0_148494_1332883_0_0_18/Final%20Report%20PA%20Financial%20Education%20Task%20Force%202013.pdf.

Public School Code of 1949—Omnibus Amendments. (2010). Act of November 17, 2010. P.L. 996. No. 104. §1551. Retrieved from http://www.legis.state.pa.us/cfdocs/legis/li/uconsCheck.cfm?yr=2010&sessInd=0&act=104.

PART II

College-Focused Financial Literacy Education

5

Raising the Bar on Efforts to Increase Financial Capability among College Students

MARY JOHNSON

Contents

As college costs continue to rise well above the rate of inflation and financial aid funding fail to keep commensurate pace with these increases, students are relying more heavily on student loans to finance their college education. Estimated at $1.16 trillion by the Federal Reserve Bank of New York, student loan debt is second only to mortgages, surpassing all other forms of consumer debt such as car loans and credit cards (Federal Reserve Bank, 2015). About seven in

10 (69%) of graduates of 4-year colleges had student loan debt in 2013 with an average balance of $28,400, according to an annual survey conducted by the Project for Student Debt, up nearly 7% from 2011 (TICAS, 2014).

Concerns over rising student loan debt levels are exacerbated by poor levels of financial literacy among young adults in America. The recent Organization for Economic Co-operation and Development (OECD) Program for International Student Assessment (PISA) Students and Money study, which measures financial literacy among 15-year-olds around the globe, for example, found that U.S. students scored an average of 492 out of a possible 700, which is below the average of the 18 participating countries/economies that took part in the study. Moreover, more than one in six did not reach a "baseline" level of financial literacy (OECD, 2014). These results may not be very encouraging, but they are not that surprising given the fact that only 17 states require high school students to take a financial literacy course in order to graduate from high school, and only six states require students to actually pass a personal finance examination (Council for Economic Education, 2014).

Without exposure to financial education early on, college students may recognize the fact that student loans represent a lot of money, but they do not fully understand the impact that this debt may have on their life after college. Recent reports have highlighted some of these negative implications, such as delaying marriage or buying a house because of unmanageable student debt. A survey conducted by the Institute of the American Institute of Certified Public Accountants, for example, found that 75% of respondents or their children have made personal or financial sacrifices because of monthly student loan payments. Forty-one percent have postponed contributions to retirement plans; 40% have delayed car purchases; 29% have put off buying a house; and 15% have postponed marriage. In addition, less than 40% reported that they fully understood the impact student loan debt would have on their future plans, and 60% felt at least some regret over funding their education with student loans (AICPA, 2013).

Many students have not had the experience of being responsible for making monthly payments or managing a budget before heading off to college, and may not understand basic financial concepts such as the difference between principal and interest. Others struggle with understanding their financial aid packages, take out more student loans than they really need to cover their direct educational costs, or do not use these funds as wisely as they should.

Student debt is even more concerning for students who take out loans and do not graduate, as they are almost four times more likely to default on their student loans than students who finish their degree (Nguyen, 2012). Studies show that finances or having to work to make more money are the major reasons students leave college before completing their degree (NCES, 2012; Public Agenda, 2011). As such, institutions of higher education are increasingly recognizing that they have a vested interest in helping students navigate not only the financial complexities of the financial aid system, but also the intricacies of day-to-day money management so that they can afford to stay in school and persist to graduation.

Financial Preparation before College

Today's college students are making major decisions about financing their college education that will have lasting implications for their future well-being, yet very little is known about the financial experience, attitudes, and behaviors of these young adults, or how this may influence the development of financial capability.

For the past three years, a national study entitled, "Money Matters on Campus," conducted by EverFi, Inc. and sponsored by Higher One, Inc., has surveyed an average of 50,000 first-year college students across the United States (EverFi, Inc., 2015). The survey asked students a variety of questions pertaining to banking, savings, credit card, and student loan usage; financial knowledge; financial attitudes; and past and planned financial behaviors to gain a better understanding of their financial readiness for college.

Financial Experience

The latest survey published in 2015 found that financial experience such as having a bank account and/or credit cards among incoming college students has increased since the survey began in 2013, but there has not been a concurrent increase in basic financial management skills or planning. Over time, students decreased the likelihood of engaging in financially responsible behaviors such as following a budget, paying credit card bills on time, and reviewing bills (Figure 5.1).

Although 62% of those attending 4-year institutions check their bank account balances regularly, 12% reported that they were too nervous to see how much money they have in their accounts. Less than 40% use a budget to help them manage their money, and only 14% use a money management app or program. This last finding is particularly surprising given the level of technological savvy among this generation, but points to the need for the development of better financial tools that are both engaging and relevant to young adults, and which meet their day-to-day needs and expectations. Long gone are the days of keeping a check register and reconciling monthly bank statements as an integral part of managing money. However, technologies such as online and mobile banking have not replaced the fundamental reason why these exercises were (and still are) so important—taking the time to examine past transactions not only for accuracy, but also to mentally review and reassess where and how money is spent. Keeping track of and knowing where your money is going is one of the most important elements of becoming financially capable, yet many young adults are not engaging in these behaviors.

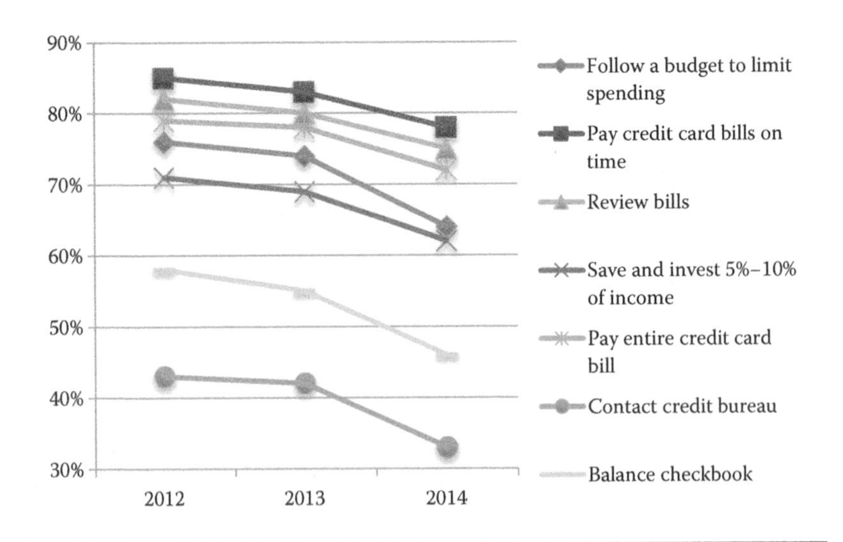

Figure 5.1 Planned financial behaviors.

The study also asked about levels of preparedness for their college experience. Surprisingly, students reported being less prepared to handle their finances than any other challenge they faced while in school such as keeping up with coursework (Figure 5.2). Those who reported having a checking account, especially an individual account, felt markedly more prepared to manage their money than those that were unbanked. This strongly suggests that increasing experiences with some type of "transactional" account before college would better prepare students to manage their personal finances when they are on their own.

In addition, the study found that even though the expected amount of outstanding loan balances has increased, students have become less likely to plan on paying them back on time and in full. This may reflect a lack of understanding of the fiscal responsibilities that come with taking out student loans. Colleges and universities would be well served to provide more assistance for students to help them stay apprised of their outstanding loan balances and estimate likely monthly payments, as well as offer more opportunities for learning about repayment and consolidation options well before graduation approaches. Indiana University, for example, instituted a new initiative in the 2012–2013 school year with a simple letter that informed

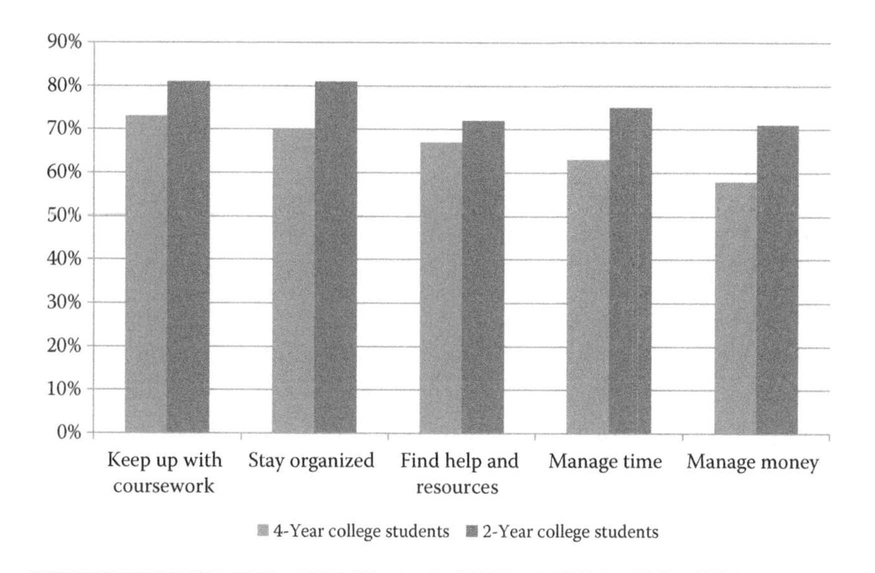

Figure 5.2 Reported levels of college preparation.

students of what their monthly loan payment would be after graduation before they signed on for additional loans. The university has reported a decrease in student borrowing of nearly $31 million from previous years (Bloomberg, 2014).

Financial Stress

The 2015 *Money Matters* investigation also considered the impact of stress on financial independence. Although about half all respondents reported being stressed about their finances, women reported significantly higher rates of stress, particularly on the issue of finding a job after graduation (78% versus 60%). In addition, Hispanic/Latino students reported the highest levels of stress across all categories. Levels of stress decline somewhat as parental education levels increase, but the strongest predictor of financial stress was the total amount of student loan debt expected upon graduation. It found only weak to negligible correlations with all financial behaviors and experience, suggesting that most students are experiencing some significant levels of financial stress regardless of prior preparation. Financial literacy programs should recognize that the development of financial capability does not take place in a vacuum (Way, 2014) and incorporate stress management strategies and interventions similar to those that have been successfully used in the public health arena to promote behavior change.

Taking Action on Campus

There is evidence that more colleges and universities are recognizing their role in helping shape a better financial future for their students, but have a long way to go in fully addressing the need. In a survey conducted in 2014 by iGrad, an industry leader in providing online financial literacy programs for colleges and universities, only about 30% of nonprofit colleges and universities reported having a financial literacy program in place at their institution (Alban, 2012).

Higher education associations also are weighing in on the importance of financial literacy on campus. For example in 2011, the American Association of State Colleges and Universities, underscored the importance of addressing this growing economic and social concern, stating in its white paper on the subject that:

> State colleges and universities have a unique opportunity to provide leadership on the critical topic by weaving financial education into the fabric of their campus communities. By delivering value-added financial literacy programs and services, college leaders can help students understand how to develop prudent financial habits. Financial education programs can also contribute to the community-based, public–purpose mission of state colleges and universities and facilitate the integration of a new generation of informed citizens into the American Economy. (Harnisch, 2010)

The paper suggests a number of recommended steps, activities, and programs to address the issue, beginning with the development of a campus-wide, coordinated financial education strategy. Such strategies should harness the resources and assets that may be already available on campus including staff from offices such as financial aid, business/bursar, and student services; faculty in such disciplines as economics, business, and finance; students and student organizations; alumni; and the local community.

The Financial Literacy Task Force of the Coalition of Higher Education Assistance Organizations, an organization of higher education institutions and servicers involved in the federal Perkins loan program, issued a white paper in 2014 on the most effective models and methods for gaining traction for financial literacy programs on campus. The paper was coauthored by this author and several leading researchers and practitioners in the field including Sonya Britt and Dottie Durband of Kansas State University, coauthors of *Student Financial Literacy: Campus-Based Program Development* (Durband and Britt, 2012), and Sharon Lechter, author of *Rich Dad, Poor Dad.* As the paper indicates, successful financial literacy efforts are those that carefully articulate campus needs; secure appropriate buy-in from key stakeholders on campus; identify a clear set of goals, objectives, and intended outcomes; and include an ongoing assessment strategy (Alban et al., 2014).

Although no singular operational model for college financial literacy programming has yet to emerge, there are several program approaches that have been gaining in popularity in recent years. These include full financial education or money management centers and peer-to-peer counseling services. Specific examples of both of these models are discussed later in this chapter. The methods for actual delivery of financial literacy education also vary widely, and include event-based programs, workshops, in-class courses, individual counseling or coaching, online courses or programs, and game-based education.

Best Practice Themes and Approaches

Since 2011, Higher One, Inc., a financial services company that provides higher education refund management, payment, and data analytic services, has provided funding to support financial literacy initiatives at colleges and universities in the United States through its Financial Literacy Counts grant program. The program offers small seed grants to help institutions begin or expand financial programs on their campuses, but requires that student representatives are involved in both the planning and execution of the programs. As engagement is one of the major hurdles in providing financial literacy education, students are well positioned to understand the challenges and stresses of everyday college life, and also bring a wealth of creativity to designing events and activities that will attract students. They also are helpful in recommending the best marketing and outreach strategies, as they are the most likely to be current with the latest social media channels and trends.

Through a review of grant reports from recent grant recipients, a number of commonalities and best practice themes are beginning to emerge.

Competency/Behavior Goals

As referenced earlier, one of the most important first steps in implementing a financial literacy program is to identify the problem(s) that need addressing, and set clear and measurable objectives for program outcomes. Grant recipients identified a variety of desired competency and behavior goals, as summarized in Table 5.1. Some focused on financial aid issues such as making sure students complete the Free Application for Federal Student Aid correctly and on time each year, or showing them how to access information on their current loan balances within the federal National Student Loan Data System for Students (NSLDS) system. Other institutions saw a more pressing need to help their students with basic budgeting, understanding credit, spending control, and identity protection—all key elements

Table 5.1 Competency/Behavioral Goals

TOPIC	GOALS
Financial aid	On-time and accurate FAFSA filing
	Understanding/keeping track of student loan balances
	How to access federal NSLDS system
	Estimating student loan payments
	Student loan repayment options
	Understanding promissory notes
	Taking advantage of other scholarship opportunities
Money management	Learning to budget
	Spending control
	Understanding credit/debt avoidance
	Earnings/income strategies
	Savings and investing
Identity protection	Online safety
Institutional goals	Retention/degree completion
	Reduction in accounts receivables and bad debt
	Student loan debt reduction
	Improved cohort loan default rates

Note: FAFSA, Free Application for Federal Student Aid.

of a comprehensive financial literacy program. From an institutional perspective, key goals identified were to increase retention and degree completion, reduce the amount of account receivable balances and/or uncollectable debt, decrease overall student loan debt burdens, and improved cohort student loan default rates.

Targeted Student Populations

Although there were large variations in both the type and number of students served by the financial literacy programming among grant recipients, first-year students (or incoming freshmen), students living in on-campus residence halls, and graduating seniors were the three specific groups of students most commonly targeted. At Eastern Washington University, for example, the housing and residential life office collaborated with the academic affairs department to offer presentations in classes that served primarily first-year students. It also offered "One-Minute Clinics" to dispense personal finance information to students as they were entering or leaving the residence halls. In another example, Missouri Valley College's "MOVAL Makes ENE about CENT$" initiative reached more than 300 students in freshman seminar courses on topics such as budgeting, savings, identify protection, and the "cost" of missing classes.

In addition to classes and freshman seminars, institutions use a variety of other opportunities to reach students at venues or events they were already attending such as welcome weeks, move-in weekends, special events, and impromptu encounters. Special events, in particular, are a great first step for institutions just beginning their financial literacy outreach efforts, as they can be relatively inexpensive to implement and generally receive positive support from the campus community. For the most part, they do not infringe on existing programs and activities. Some institutions have reported that getting support to embed a financial literacy component into an existing program such as a first-year experience course or freshman orientation schedule that is "owned" by another office or department can sometimes be difficult, as those agendas are usually considered full or competing in theme. Similar resistance is common when considering a requirement or mandate for a financial literacy program or course. This highlights the importance of including critical stakeholders from all aspects of campus life as part of the financial literacy planning efforts—the need for campus buy-in to ensure the program's success and continuation cannot be overstated.

At the University of Wisconsin–Oshkosh, a team from the financial aid and career services offices worked collaboratively to secure a 1-hour time slot during move-in weekend on campus. They developed a "Money Madness" game modeled after the popular game show, "Jeopardy," which was offered at a variety of locations on campus on the same day by a group of student leaders, reaching 1750 students campus-wide. Other examples of special events programs include grocery bingo at Iowa Western and a scavenger hunt at Moreno Valley College. Again, involving students can help increase the level of creativity and impact. It should be noted, however, that special events tend to be short in duration and not in-depth enough to have truly measurable long-term effects on student behavior. That said, they do provide important "mental moments" for students to reflect on their attitudes, spending habits, and other financial goals, and help to bring visible attention to the issue on campus.

More impromptu encounters involving students are also gaining in popularity, particularly with the growth and interest in short video production and online dissemination. Students at the University of Minnesota and Missouri Valley College, for example, roamed the campus as "news reporters," interviewing students and asking them a variety of financial literacy-related questions such as, "Do you know the interest rate on your student loans?" These encounters were recorded and posted to financial literacy websites and Facebook pages, and/or shared via other social media channels. Students at Iowa Western Community College scripted, shot, and edited four "Money Minute" financial literacy videos that featured students in real-life situations finding out about credit cards, budgeting, borrowing, and student loan repayment.

Facilitation and Promotion

As mentioned above, a major challenge of achieving success with any financial literacy effort on campus is getting students to actually attend workshops, classes, or events, and engage in those activities. Without an explicit mandate, simply putting resources up on a website or offering a workshop in the cafeteria is unlikely to garner much interest or attendance unless facilitation and promotional strategies are carefully planned out, tested, and executed.

The most successful promotional strategies are those that begin with having a recognizable or appealing "brand" that can be used on websites, in social media channels, emails, posters, banners, and the like. Some examples are provided below:

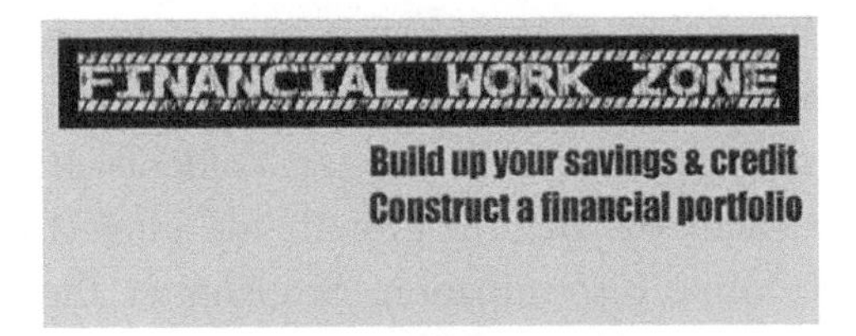

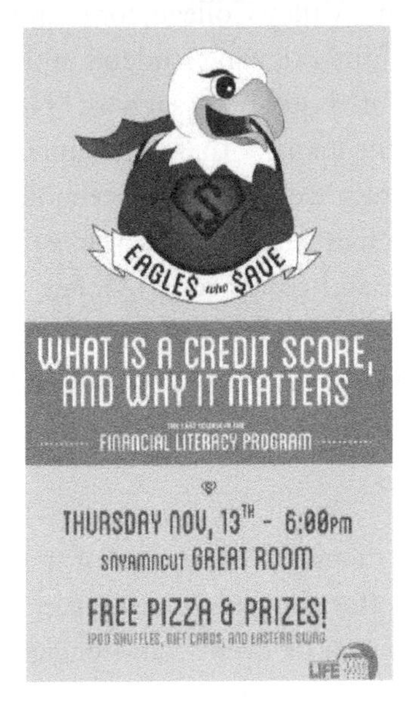

Other strategies to attract students include providing food and other refreshments; incentives such as gift cards or book store scholarships, and extra course credit for attending the workshop. For community college students who tend to be commuters and do not spend as much time on campus as their 4-year counterparts, working with faculty to provide incentives or incorporate financial literacy lessons into the classroom may be key to increasing attendance.

Peer-to-Peer Counseling

Another effective approach is peer-to-peer counseling or coaching. The basic premise of this model is that the best way to reach students is through other students who can relate to the daily issues and struggles that students face. In most programs, peer mentors or coaches undergo some kind of personal finance training, usually by staff certified in financial education and/or counseling. These mentors then provide in-person or telephone appointments, group presentations in classes, student clubs, residence halls, and/or plan and host special events. Some institutions that have more robust financial literacy offerings include a peer-to-peer counseling component as part of their service offerings. Some examples of peer-to-peer programs include those at Kansas State University, Ohio State University, Armstrong State University, University of North Dakota, and Emory University.

Some institutions are taking steps to reach students before they actually get to campus by having peer mentors and/or staff offer workshops to local high schools. A relatively new and exciting initiative is the nonprofit organization called "Moneythink" (see Chapter 3), self-described as a movement of young people dedicated to restoring the economic health of the United States through financial education. The program, which began in 2009 on the campus of the University of Chicago, places trained college student volunteers in urban high schools to service as peer models and financial mentors for teenagers. Recognized by President Obama as one of the most promising social innovations in the nation, it now has 30 chapters in 10 states, and continues to grow. It also developed a new Moneythink Mobile app, dubbed the gamified Instagram for finances. Students use the app to complete digital challenges and earn points and social reinforcement outside the classroom to build financial awareness and habits outside the classroom. Several higher education institutions are also testing the efficacy of using the app with first-time college students.

Online Programs

There are a growing number of online financial literacy programs designed specifically for the college student population. The advantages of using an online platform are the potential to reach large numbers of students, and the relatively inexpensive cost per student (as compared to using full-time staff or adding on-ground courses). In addition, program content tends to be more interactive and appealing to students. The National Endowment for Financial Education (NEFE), for example, recently updated its CashCourse program, which is freely available for schools that register (see Chapter 6). According to the information provided on the NEFE website, about 900 colleges and universities have signed up to use the course. Colorado College also offers a free online program, as well as resources and materials for educators. In addition, there are a number of third-party providers of interactive, online programs including EverFi, iGrad, and SaltMoney.

Online programs alone are not the panacea for solving the financial literacy gap, as institutions that use these programs face similar challenges with student engagement and usage as they would with other, more traditional methods of delivery. Simply providing a link on a website or emailing students to let them know about the availability of the program will not guarantee students will use the program. The most success when using an online platform comes when there is a corresponding mandate or incentive for course completion and/or continuation. In addition, although these programs have been shown to effectively boost short-term knowledge gains and, in some cases, behavior changes, their effectiveness on longer term development of financial capability and behavioral change has yet to be documented.

Assessment and Outcome Measurement

Securing support and resources for a sustained financial literacy program on campus can be challenging given other competing needs and interests on campus, so it is really important to assess and communicate the impact of any financial literacy efforts, whatever their length or intensity. College and university administrators want evidence that what they are supporting is actually working; and practitioners need continuous feedback to know how to improve programs and to identify new or changing needs and priorities. With the key goals of increasing financial knowledge and responsible financial behavior, providing pre- and postprogram surveys, quizzes, and tests are the most common instruments used, and provide the clearest indication of short-term benefits. Gauging student satisfaction with programs and services also should be part of any assessment strategy, particularly with regard to program content and delivery method.

There are a few institutions (e.g., Kansas State University and The Ohio State University) that are working to measure the longer term impact of programs on knowledge, satisfaction, and stress. In addition, some online programs include follow-up surveys several months after the intervention. Perhaps more promising, however, is the University of Arizona's APLUS (Arizona Pathways to Life Success) study—one of the first longitudinal studies of young adults' changing financial knowledge and practices. The study began by surveying a cohort of first-year students in 2007, just before the Great Recession, and continues to resurvey these young adults over time into middle age. The latest "Wave 3" report findings support the positive and cumulative impacts of early and repeated financial education for young adults. The study also points to the benefits of working to developing positive financial behaviors early on as a pathway to adult "self-sufficiency" (Serido and Shim, 2014).

Institutional Showcases

There are a number of strong and viable financial literacy programs that provide a combination of all or most of the elements included in this chapter. Three of those institutions are highlighted here to provide more detail about their focus and service offerings.

Sam Houston State University (Texas)

Sam Houston State University Money Management Center (see Chapter 9) is a financial outreach and education program within the division of student services established in 2008. Its mission is to empower students with unbiased financial literacy education and tools to help them achieve financial independence. The impetus for the creation of the center goes back to observations with the university's carding office that students were asking about or using their financial aid refunds not for educational expenses such as books, but for spending totally unrelated to their college education including shopping, entertainment, and travel. (Financial aid refunds are funds that are left over from students' financial aid awards after all "direct" tuition and fees have been paid.)

As specified on its website, the Center's goals, among other things, are to help students:

- Learn how to track expenses and create a personal budget.
- Create a financial plan for college.
- Graduate with a plan for repaying student loans.
- Understand employee benefits and retirement plans before deciding on a career opportunity.

All staff members of the Center are certified personal financial counselors. It provides a variety of services including financial coaching sessions, workshops, special events, and classroom presentations.

Ohio State University

The Ohio State University takes a more holistic approach to incorporating financial literacy education to its students (see Chapter 10). Housed within its student wellness center, its Scarlet and Gray Financial program provides financial education to undergraduate, graduate, and professional students. It offers free, confidential one-on-one peer financial coaching provided by trained student mentors, as well as educational presentations to students in first-year experience courses, campus organizations, residence halls, and other courses. Its stated mission is to empower students "to create the life they desire through the use of a goal-driven process that encourages the development of health attitudes and financial behaviors." The specific curricula areas it focused on include financial goal setting, the basics of banking, budgeting, credit education, and debt repayment education.

University of Illinois

The Student Money Management Center at the University of Illinois seeks to empower students to make positive behavioral changes. It uses a variety of outreach methods including a robust website, peer-to-peer mentoring, webinars, workshops, and other presentations. It also uses the iGrad financial literacy platform, which offers informative articles, interactive budgeting tools, engaging webinars, and an online community where students and staff across the United States can connect on common financial topics.

More recently, the center launched a "Financial Literacy Badges Program," which utilizes an online platform for students to earn and display financial accomplishments around five core financial competencies: earning, spending, saving, borrowing, and protecting. Students are required to complete at least three educational activities in each competency area to earn a badge. One interesting component of the program requires students who received a financial aid refund to take a student account refund quiz to remind students that these funds should be used wisely. Almost 3000 students participated in the program in fall 2014.

Summary

Colleges and universities are increasingly recognizing that they have a vested interest in helping students to borrow responsibly and effectively manage their finances while in school, as these are key factors in successful degree completion and financial stability after college. Recent studies are beginning to shed light on the complexity of financial literacy development, particularly the extent to which financial experience, attitudes, and stress can impact financial behavior. Providing more financial literacy education opportunities earlier on, particularly in high school, would help to better prepare students for their college experience. But until more states incorporate financial literacy education into the K–12 core curriculum, colleges and universities must help to fill the gap.

College financial literacy programs are increasing in number and intensity, with programs ranging from special events and workshops, to full-service money management centers. Peer-to-peer mentoring models also are shown to be effective in reaching and engaging students. However, competing resource limitations constrain the growth and sustainability of these initiatives. As no distinct model for college financial literacy programming has yet to emerge, student involvement in design and delivery has been shown to increase student participation and engagement. In addition, incorporating strategies that focus on changing financial attitudes and reducing stress may enhance program outcomes.

As outreach efforts continue to grow, more research and assessment on the efficacy and long-terms impacts of financial literacy interventions is needed to more fully understand how financial capability develops before and during the college experience, and what interventions produce the most positive and lasting results.

References

Alban, C. (2012). College Financial Literacy Compendium. Retrieved on April 28, 2015 from http://schools.igrad.com/financial-literacy-resources/.

Alban, C., Britt, S., Durband, D., Johnson, M. K., and Lechter, S. (2014). Financial Literacy in Higher Education: The Most Successful Models and Methods for Gaining Traction. Retrieved April 28, 2015 from http://coheao.com/wp-content/uploads/2014/03/2014-COHEAO-Financial-Literacy-Whitepaper.pdf.

American Institute of CPAs (2013). Realities and Regrets of Student Loan Debt. Retrieved on April 5, 2015 from http://www.aicpa.org/press/pressreleases/2013/pages/aicpa-survey-reveals-effects-regrets-student-loan-debt.aspx.

Bloomberg Business (2014). *How Students at a U.S. University Borrowed $31 Million Less.* Retrieved on April 28, 2015 from http://www.bloomberg.com/news/articles/2014-07-03/here-s-how-indiana-university-students-borrowed-31-million-less.

Council for Economic Education (2014). *Survey of the States 2014.* Retrieved on April 1, 2015 from http://www.councilforeconed.org/policy-and-advocacy/survey-of-the-states/.

Durband, D. and Britt, S. (Eds.). (2012). *Student Financial Literacy: Campus-Based Program Development.* New York: Springer.

EverFi, Inc. (2015). Money Matters on Campus: How College Students Behave Financially and Plan for the Future. Retrieved on April 10, 2015 from http://moneymattersoncampus.org/.

Federal Reserve Bank of New York (2015). Quarterly Report on Household Debt and Credit. Retrieved on April 28, 2015 from http://www.newyorkfed.org/householdcredit/2014-q4/data/pdf/HHDC_2014Q4.pdf.

Harnisch, T. (2010). *Boosting Financial Literacy in America: A Role for State Colleges and Universities.* New York: American Association of State Colleges and Universities.

National Center for Education Statistics (2012). Higher Education: Gaps in Access and Persistence Study. Retrieved on April 14, 2014 at http://nces.ed.gov/pubs2012/2012046/tables/e-38-1.asp.

Nguyen, M. (2012). *Degreeless in Debt: What Happens to Borrowers Who Drop Out. Charts You Can Trust.* Education Sector.

OECD (2014). PISA 2012 Results: Students and Money (Volume VI). Retrieved on April 28, 2015 at http://www.oecd.org/pisa/keyfindings/pisa-2012-results-volume-vi.htm.

Public Agenda (2011). With Their Whole Lives Ahead of Them. Retrieved on April 15, 2015 at http://www.publicagenda.org/files/theirwholelivesaheadofthem.pdf.

Serido, J. and Shim, S. (2014). *Life After College: Drivers for Young Adult Success.* Tucson, AZ: University of Arizona.

The Institute for College Access and Success (2014). Student Debt and the Class of 2013. Retrieved on April 15, 2015 at http://ticas.org/posd/home.

Way, W. L. (2014). Contextual Influences on Financial Behaviors: A Proposed Model for Adult Financial Literacy Education. *New Directions for Adult and Continuing Education*, 2014(141), 25–35.

6

Financial Literacy

National Endowment for Financial Education

AMY MARTY

Contents

For the past 30 years, the National Endowment for Financial Education (NEFE) has been providing personal financial literacy materials to a wide variety of audiences, including students, educators, and consumers. This chapter covers some of NEFE's recent work in higher education and financial literacy.

I am a program manager at NEFE. I oversee an online college financial literacy program called CashCourse, which as of this publication is used at about 900 colleges and universities in the United States. CashCourse will be discussed in more detail shortly.

Overview of NEFE

NEFE is a private nonprofit foundation and 501(c)(3) organization based in Denver, Colorado. It is independently funded, noncommercial, and politically neutral, and offers all of its materials and services to the public at no cost.

NEFE began in 1972 as the Denver-based nonprofit College for Financial Planning, which created the Certified Financial Planner certification and helped establish financial planning as a profession. NEFE was created in 1992 as the parent entity of the College, and in 1997 the College was sold to an outside organization, with proceeds from the sale going to fund NEFE's current endowment. NEFE does not accept financial support from product sales, individuals, government, or corporations; financial growth comes from investment of the endowment's assets. The first NEFE financial education program, the High School Financial Planning Program, was created in 1984 and is used at more than 5000 high schools and other educational organizations to this day.

In the field of financial education, there are competing commercial interests, government interests, consumer concerns, and nonprofit players in the field, all attempting to engage the American public and influence behavior. NEFE's role has been as a provider of research and educational materials for educators and consumers, without any political or commercial ties influencing the materials we provide.

Financial Literacy in Higher Education

NEFE established CashCourse—its financial literacy program for colleges and universities—in 2007, piloting the program with alumni associations in the Big Ten conference. CashCourse was created after seeing a gap in financial education for post–high school young adults, based on demand for a credible, scalable program, as well as mounting concern over student debt. At the college level, flexibility is needed in a financial literacy program, as each campus and its students are unique; creating an in-the-box "course" might have limited accessibility because of the varying needs and course acceptance criteria at each school. Therefore, CashCourse was conceived as a web-based financial resource library that addressed the concerns of undergraduate students and the instructors and administrators who work with them.

Although web-based and accessible by students on their own time, CashCourse relies on support from faculty or staff champions—individuals who take it upon themselves to use the program and its resources with students. One of the challenges for the growing number of online financial literacy programs is the expectation that a given course will work perfectly for a campus right out of the box, especially as an optional activity. In reality, given the option, students typically will not opt-in to financial education, unless those students are already somewhat financially capable and proactive in managing their money. For this reason, the more that campuses can integrate financial literacy into mandatory or popular programming, the more likely it is that students will be receptive to the concept of personal finance and engage with the materials. It's not as if students don't see the value of the material; rather, the natural inclination for many students is to ignore their finances until a crisis or emergency prompts them to take action. Driving students to seek out financial information on their own remains a challenge.

Student loans have been a popular topic of discussion within the field of financial education, as well as with the general public and the media. The conversation around smart borrowing is an important one to have, but it often overshadows other topics of concern, especially with emerging households. Creating a standard definition of what makes a financially capable person—beyond the ability to repay loans—is critical in determining what areas of focus deserve the attention of educators and institutions of higher education.

Research: Young Adults and Financial Literacy

Recent research funded by NEFE gives insight into the challenges of helping young adults become financially capable. This chapter gives a brief summary of the research, and the full reports can be found in the research section of NEFE's website.

In June 2013, NEFE published research by Dr. John G. Lynch, Dr. Daniel Fernandes, and Dr. Richard G. Netemeyer titled "The Effect of Financial Literacy and Financial Education on Downstream Financial Behavior." The study was a scientific review and meta-analysis of existing financial education research exploring "the link between financial education, literacy, and behaviors" (Lynch, Fernandes, and Netemeyer, 2013). The study gives data to support the generally accepted idea that financial literacy drives better decision-making. Key findings are included below:

1. The amount and timing of financial education matters. When it comes to attempts at building financial literacy to shape behavior, education that closely precedes a financial decision has more impact.
2. Behaviors and literacy as measured to date are weakly linked. Educational interventions and financial literacy as measured to date are only weakly linked to behaviors. Moreover, in studies that measured financial education effects on both knowledge gains and behavior, effects of financial education delivered through interventions were much less than education in comparable domains, such as workplace education or career counseling.
3. Findings from past investigations merit revisiting. Different types of studies have yielded such disparate results—more varied than science would predict—that we must question to what extent those differences stem from widely varying research designs and analyses (Lynch, Fernandes, and Netemeyer, 2013).

In terms of the amount and timing of financial literacy, Figure 6.1 shows the size of the impact on decision-making relative to the length of time of each financial literacy "intervention" over time. For example, a 24-hour class right before a financial decision was much more likely to have an impact than that same class 2 years before the decision, or a much shorter class a few months before the decision.

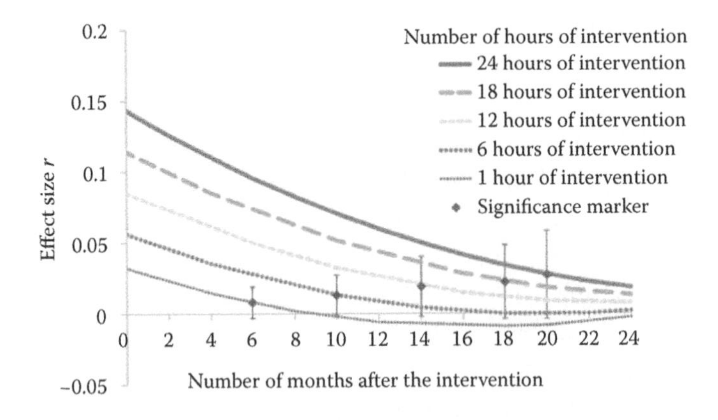

Figure 6.1 Case for timely financial education.

The study concluded that researchers need to standardize and contextualize their reporting on financial education, "clearly identifying changes in knowledge, behavior, or both" (Lynch, Fernandes, and Netemeyer, 2013), and that research should explore which interventions work best with which audiences, and how marketing strategies and campaigns affect financial behavior and financial literacy.

For college students and young adults, recent NEFE-funded research highlights how college and consumer debt impact attitudes and financial choices after graduation. Research study by Dr. Randy Hodson and Dr. Rachel E. Dwyer from the Department of Sociology at Ohio State University titled, "Financial Behavior, Debt, and Early Life Transitions: Insights from the National Longitudinal Survey of Youth, 1997 Cohort," finds that "debt facilitates achievement but can lead to emotional or financial strains and delay family formation" for early Millennials (Hodson and Dwyer, 2014). Some key findings of the research are included below:

- Compared to Generation X, Millennials took on greater amounts of credit card debt at an earlier age.
- Millennials at 4-year colleges who also had student loans were more likely to delay marriage; the same was true for millennial men at 2-year colleges.
- By 2009, 50% of millennial homeowners were under water on their mortgages. In preceding years, this number was less than 10%.
- Credit card debt among Millennials increased anxiety and depression and most affected lower-income households.
- In the wake of the Great Recession, Millennials across all income levels reported financial strain and trouble making ends meet.
- Millennials who got financial advice from their parents reported less financial strain than peers who received advice from nonprofessionals or who received no advice (Hodson and Dwyer, 2014).

This research gives some interesting insight into the lives of early Millennials and their finances, but it also raises questions about how the financial literacy field should address stages of transition in early adulthood.

Longitudinal research of college graduates, based on research out of the University of Arizona, has analyzed how young adults develop financial knowledge and how they transition into adulthood, especially given the economic environment they encountered during the recession. The NEFE-funded study, entitled "Arizona Pathways to Life Success for University Students" (also known as the APLUS study), is led by Dr. Joyce Serido of the University of Minnesota and Dr. Soyeon Shim of the University of Wisconsin–Madison. Wave 3 of the research was released in 2014.

APLUS follows a group of young people who began college at the University of Arizona in the fall of 2007, making the participants in their mid-twenties during the third wave of the research. This post-college wave of the study investigates how the participants approach the beginning of their careers and the transition to adult life.

More than half of the participants—including nearly half of those employed full time—relied on at least some financial support from their families, and this demand puts their parents' retirement planning at risk. Participants faced challenges in paying off student loans, buying a home, reaching savings goals, and making major purchases because of their financial instability (whether actual or perceived instability). Participants' financial instability impacted their life satisfaction and psychological and physical well-being. Income and debt level further impact quality of life, and "for participants with debt, financial well-being was 17% lower, 19% lower and 31% lower for those employed full-time, employed part-time and unemployed, respectively. Debt was associated with 4%, 8%, and 10% lower life satisfaction for those same groups" (Serido and Shim, 2014).

Finally, financial behaviors practiced while in college have an impact on early adulthood. The study finds three pathways:

> High-functioning participants (12%) maintained consistently high levels of responsible financial behavior across all three waves; Rebounding participants (61%) started college with moderately responsible financial behaviors that had declined by year four but rebounded by Wave 3 two years later; Struggling participants (26%) started college with poor financial behaviors, which had further declined by year four; though their behaviors had improved two years on, they were still worse than during their first year of college and significantly lower than all other participants. (Serido and Shim, 2014)

For higher education providers, the APLUS research reinforces the importance of financial education in school on students' well-being after graduation, but also reveals that attitudes and parental involvement before college are vital. In summary, "Better financial attitudes, higher parental expectations, and having financial education promoted a more successful transition to young adulthood; a lack of financial education, lower perceived financial knowledge, less perceived financial control and lower parental role modeling made for a less successful transition" (Serido and Shim, 2014).

Five Key Factors for Effective Financial Education

In 2014, NEFE's director of education, Dr. Billy J. Hensley, developed a set of principles for financial educators, the Five Key Factors for Effective Financial Education. These standards encourage a minimum standard of what is considered financial education and are intended to lead to more consistency in the field. The guidelines are included below:

1. **Well-Trained Educator** (and/or tested e-learning protocol). Simply integrating personal finance topics into a learning environment is not enough. The educator needs to be confident, competent, and knowledgeable about the topic of personal finance in order to create a learning environment that is ideal for student learning. Educators attain this proficiency by attending college-level courses and/or post-certification workshops that have been evaluated to demonstrate impact in increasing instructor effectiveness. Fundamentally, educators should demonstrate high levels of understanding—both with the content and the pedagogy—of the topics that espouse the tenets of financial capability.
2. **Vetted/Evaluated Program Materials.** The content and program materials (e.g., classroom activities, topics, examples, and assignments) should be created with the consultation of field experts (e.g., insurance agents and financial planners) and tested to be appropriate for the audience for which instruction is being implemented. For example, all instructional materials should include correct and up-to-date information, be guided

by complex learning outcomes and objectives that are age appropriate, and tested to be impactful by external evaluators.

3. **Timely Instruction.** Program goals, instructional tools, and instruction topics should link to decisions that learners are readily able to make. If the topics cover concepts that are many years away from the capability of those participating in the instruction, alternative examples should be instituted that convey similar concepts, but that can be relatable to a near-term decision or implementation. This concept is especially true if the program has a limited time of exposure to convey the content. For instance, using a considerable amount of time focusing on mortgages, when the learners are 16, may be less effective than covering borrowing (using examples such as student loans or automobiles) and highlighting the planning process associated with attaining secured debt. In addition, learners should have access to program materials beyond any formal instruction to allow the opportunity for utilization of content and exercises at times that are opportune.
4. **Relevant Subject Matter.** As with timely instruction, relevant subject matter is essential in not only engagement with the content, but also in the prospect of impacting behavior. If learners are unable to relate to the topics, examples, and content, then the level of engagement the instructor seeks will not be achieved. Consider young adult learners who are attending a free community money management workshop who mostly have jobs that do not offer 401(k) or 403(b) savings plans. Without an understanding of the audience and the context, an instructor may focus their presentation examples of saving on the need for diversified mutual funds instead of conveying the same concepts by discussing saving, having an emergency plan, and explaining the difference between savings accounts, certificates of deposit, and Roth IRAs (Individual Retirement Accounts).
5. **Evidence of Impact (Evaluation).** Continuously seek information on the impact of a program. Well-designed evaluations, which can be managed internally or by an external party, tell educators where they are having impact on behavior,

knowledge, and/or confidence, where students are engaged, and where improvements need to be made. Without evaluation, instructors rely on anecdotes to inform their work, where a more robust assessment can show where immediate improvements can be made and which areas of success can be capitalized (National Endowment for Financial Education, 2014).

Lessons from CashCourse

Based in part on the Five Key Factors, and from qualitative research gathered during interviews with participating schools, NEFE staff have noticed some trends that indicate successful financial literacy programs at the college and university level.

First, there is no "one size fits all" approach to financial literacy. Each campus we work with is very different, and although there are lessons to be learned from peer institutions, it often takes several semesters for a school to find the exact type of programming that works for its student body. Often, a staff member or department will decide that they want to offer financial education to their students, then they come up with ideas on how to approach the task, but get discouraged when student reception is lacking. Being able to go through several rounds of trial and error can ultimately lead to a stronger financial literacy program.

Creating personal takeaways during each phase of student involvement is also critical. Financial education can be immediately useful to people who take a class, visit during office hours, read an article online, and so on. When given the opportunity to work with students, the intervention should include a personal takeaway, whether that takeaway is a savings plan, debt repayment strategy, checklist of documents needed for tax preparation, or whatever makes most sense given the content of a lesson. Instead of teaching financial lessons with theory, give students the opportunity to practice what they have learned and put the knowledge to immediate use. This echoes the research from Lynch, Fernandes, and Netemeyer about the timing of financial education in relation to decision-making. The ability to practice and apply financial education in a student's real life adds value to the student's experience at their college or university.

One strategy that NEFE has seen successful schools use is making personal finance a required part of the student's college experience, or otherwise incentivizing students to engage with personal finance. When this material is part of a required course (e.g., freshman seminar or first-year experience), it is obvious that the first barrier to building student interest is removed. Other campuses tie personal finance programming to receiving school scholarships or grants, academic advising, satisfactory academic progress, and athletics. However, if creating required personal finance programming is not an option on a given campus, then the resulting opt-in financial activities need to be very low involvement, with a low barrier to entry for students. Often, these optional activities attract students who are already proactive in managing their finances, while excluding other students.

Campuses that use peer mentors or counselors report more perceived success in their financial education efforts, although the reasons for this success are unclear. One possibility is that creating opportunities for students to talk to one another about their financial concerns works toward a more open campus culture when it comes to personal finance. Students may also feel more comfortable getting help from a peer than from a faculty or staff member. If program visibility is an issue, partnering with student organizations to provide financial literacy to their membership has also proven to be an effective way to boost participation.

As indicated by Lynch, Fernandes, and Netemeyer's research, financial education is not intended to be a one-and-done topic for students. Campuses that create lots of opportunities for students to build financial skills tend to be more successful overall with engaging students. Designing programs that are scalable and applicable to a variety of settings benefit these campuses by sharing resources and staff time among departments. Each campus area is able to reach students in different ways at different times; collaboration not only takes the burden off of a single department, but also increases program effectiveness.

Finally, measurement and evaluation are key parts of financial literacy at successful colleges and universities. Although there is discussion in the financial education field as to what metrics to pay attention to, campuses that select a few measureable outcomes and compare student performance against those criteria tend to be more effective.

Cohort default rates of student loan borrowers are a common measurement, although they do not measure many other important aspects of students' financial capability. Although there is no standard measurement for the success of a financial literacy program, colleges and universities that select a few key markers of students' financial health are better able to see where their efforts are best applied.

System-Wide Financial Literacy

In December 2013, NEFE was approached by the California Community College Chancellor's Office (CCCCO) on a default prevention plan for the system's campuses with the highest student loan default rates. As part of this project, the CCCCO also wanted to roll out a financial literacy initiative to each of its 2.1 million students at 112 colleges.

After several months of discussion, multiple presentations to the CCCCO and faculty jury, and several question-and-answer sessions, NEFE's college financial literacy program CashCourse was selected as the financial literacy partner of choice for the California Community College system.

While developing a strategic vision and plan for the program, NEFE and the CCCCO have focused on three major areas: integration of CashCourse into campus programming at every college in the system; establishing a campus culture of informed financial behaviors on participating campuses; and increasing financial capability for both students in the system and communities throughout California.

NEFE and the CCCCO facilitated the creation of a strategic plan for the first 5 years of the project. The plan details stakeholder concerns, areas of implementation for financial literacy, and outreach to students beyond the classroom, as well as further defining NEFE's involvement via CashCourse.

Phase one of the strategic plan involves establishing CashCourse as the provider of choice on all California Community College campuses, and additional phases will involve community outreach, classroom implementation, and possibly a form of instructor training.

NEFE also added a student ID number to California Community College student accounts on CashCourse in order to cross-reference a student's interaction with the program with any institutional records.

These data will be helpful down the road when assessing the impact of participation with financial literacy materials on measures such as loan default, borrowing levels, and persistence toward a degree.

This is a very exciting opportunity for NEFE to explore a more in-depth utilization of CashCourse in a community college setting, as well as to analyze the effectiveness of CashCourse on students' knowledge and behavior over a 3- to 5-year period. We see the partnership with the CCCCO as a pilot program for use of CashCourse within a state system, and the information gathered from this partnership can be used in future scenarios or systems across the country.

Final Thoughts

Financial education has become a more high-profile topic in recent years, thanks in part to the 2008 recession and the growing amount of student loan debt in the United States. Institutions of higher learning can provide a positive influence on students and prepare them for their financial lives after graduation, but engaging students will always pose a challenge to educators and staff members. Adapting financial education strategies based on research can help further the goals of colleges and universities and prepare young adults for their postgraduation financial lives.

References

Hodson, R., and Dwyer, R. (2014). Money and Milestones: The Impact of Debt on Young Adults' Financial Life Transitions. In *Financial Behavior, Debt, and Early Life Transitions: Insights from the National Longitudinal Survey of Youth, 1997 Cohort*. National Endowment for Financial Education.

Lynch, J., Fernandes, D., and Netemeyer, R. (2013). Examining Financial Education: How Literacy and Interventions Affect Financial Behaviors. In *The Effect of Financial Literacy and Financial Education on Downstream Financial Behaviors*. National Endowment for Financial Education.

National Endowment for Financial Education. (2014). Five Key Factors for Effective Financial Education.

Serido, J., and Shim, S. (2014). Executive Summary. In *Life After College: Drivers for Young Adult Success, APLUS Arizona Pathways to Life Success for University Students Wave 3*. National Endowment for Financial Education & Citi Foundation.

7

Financial Literacy

A Pathway to Financial Well-Being

THEODORE R. DANIELS

Contents

Society for Financial Education and Professional Development Financial Literacy Education Initiative

The Society for Financial Education and Professional Development, Inc.'s (SFE&PD) mission is to enhance the level of financial and economic literacy of Americans in communities throughout the nation. The thrust of our national financial literacy seminars and workshops, augmented by a newsletter, webinars, and social media, is to enable individuals to gain a better understanding of personal financial management concepts and their practical applications. These seminars/workshops represent the core initiative of SFE&PD's financial educational programs, which are specifically tailored to reach college students and include an understanding of consumer protection laws and regulations.

Research shows financial literacy programs that guide students and their families are a critical factor for college retention. This is the reason why the Society developed customized financial education programs for students to help them manage their finances wisely. College and graduate students, especially those from low-income, first-generation families, benefit greatly from a support system to help them navigate college and understand financial responsibilities.

Through our seminars, students learn practical financial strategies necessary for everyday life to make informed decisions regarding credit, spending, saving, investing, retirement, and estate planning. Financial literacy also prepares young people to become employable by teaching them how to responsibly manage their credit. To date, SFE&PD has presented financial literacy training to more than 200,000 individuals at more than 90 colleges and universities nationwide; more than 70 of those educational institutions were at historically black colleges and universities (HBCUs).

The Society envisions a nation where every individual is taught personal money management concepts and their applications in the use of financial resources. A national commitment to teaching financial education would empower all Americans to have more successful financial outcomes for themselves and their families as well as participate more fully in a broader spectrum of the American economy.

A number of leading financial sources, including research papers, studies, and surveys, indicate that individuals and households whose level of financial literacy is highest have the greatest amount of wealth and economic security. For example, the differences in asset ownership among various groups in the United States indicates a variation in preferences for saving and consumption, which could help explain the disparity in financial management and behavior of minorities.

The Federal Reserve Board's *Report on the Economic Well-Being of U.S. Households* provides a snapshot of the self-perceived financial and economic well-being of U.S. households and the issues they face, based on responses to the Board's 2013 *Survey of Household Economics and Decisionmaking*. The report indicates that experience and expectations with credit appear to vary by race and ethnicity. However, the report goes on to state that the effect is partially explained by other factors correlated with race/ethnicity and credit, such as education levels (Schmeiser, 2013). Moreover, we have found that throughout the world a common consensus exists that populations with the lowest level of financial literacy have the lowest level of financial assets and limited economic security.

The infusion of financial literacy into a population offers an opportunity for individuals to gain the financial knowledge needed to become economically secure and maintain households that are financially stable. Moreover, economic security can eliminate the financial stress that leads to other socioeconomic problems, which have a tremendous impact on society. This statement is further amplified by a recent study indicating that financial stress leads to bad financial decision-making in the use of available financial resources and to other matters that are detrimental to one's financial well-being or economic security.

Each day, individuals make financial decisions consciously or unconsciously. How they make those decisions can have a tremendous impact on their ability to become economically secure. The endgame of each person's labor is economic security. Therefore, it is important that each individual gain the financial knowledge and skills needed to make informed and sound financial decisions regarding the use of his or her financial resources. Such decision-making also includes an understanding of how much of their financial resources should be committed to business or financial transactions with limited value or growth.

The primary reason to increase one's financial literacy is to be prepared to make informed and sound decisions regarding the use of credit, establishing and accomplishing financial goals, saving and investing, risk management (insurance), retirement and tax planning. Furthermore, it is important to understand how each of these key components of personal finance interrelates. For example, the mismanagement of credit can limit one's ability to acquire economic security, drain financial resources from a household, and limit one's ability to accumulate retirement savings and employment opportunities.

A growing number of studies and surveys point to the need to increase the financial literacy of African Americans. For example, the Federal Deposit Insurance Corporation's survey of "Unbanked and Underbanked Households" indicates a high percentage of African Americans who do not have a banking relationship with a financial institution (Ryan, 2009). Surveys and analysis of data also show African Americans have low savings rates, utilize predatory lenders, mismanage credit, maintain low levels of insurance to protect their income or assets, have insufficient accumulation of retirement savings, have a 44% homeownership rate, have low median wealth, and have high debt levels. In fact, many African Americans don't realize how adverse credit ratings can ruin their opportunities for employment, insurance, checking accounts, and a host of related needs that impact their quality of life.

The Federal Reserve Bank of St. Louis, Missouri, Center of Household Stability's 2013 Annual Report, titled "Rebuilding Family Balance Sheets, Rebuilding the Economy," indicates that although all Americans lost wealth because of the Great Recession, younger and less-educated African Americans and Hispanic families lost the most. (Wealth is defined as the total family assets minus liabilities or money owed. If the difference is positive, the family has wealth, and conversely, if the difference is negative, the family does not have wealth.)

The report goes on to state that a growing body of research shows that healthy balance sheets, and not just income, matter for basic household stability (Boshara, 2013). Hence, it is important to use income wisely to increase financial stability and overall financial well-being throughout one's lifetime. This is an area where many African Americans have not been successful. Many use their income for consumer consumption and as a result, limit their savings. In fact, many withdraw funds from retirement savings accounts to cover their children's college cost or to make ends meet. Such withdrawals result in a loss of future wealth. However, a significant number of African American individuals and families do invest a portion of their income and make investments, but such investments are conservative—for example, certificate of deposits, savings accounts, and bonds. These investments offer limited opportunities for wealth creation or an increase in net worth. Whites, however, have more diverse investments, such as stocks and bonds, which offer a greater opportunity for wealth creation for financial well-being.

In February 2015, the Center for Household Financial Stability at the Federal Reserve Bank of St. Louis, Missouri, issued the first in a series of essays titled "Demographics of Wealth." The essay is the result of an analysis of data collected between 1989 and 2013 through the Federal Reserve's Survey of Consumer Finances. The essay makes a connection between race or ethnicity of an American family's wealth level and how its financial affairs are managed. It states that race and ethnicity are strongly associated with financial behavior and outcomes.

For instance, one key finding indicates that when looking at median family wealth (assets minus liabilities) the ranking of the four racial groups (White families, Asian families, Hispanic families, and African American families) did not change order between 1989 and 2013. White families ranked first, followed by Asian families, Hispanic families, and African American families. The median wealth levels of Hispanic and African American families are about 90% lower than the median wealth of White families, yet median income levels of Hispanics and African Americans are only 40% lower (Boshara, 2015).

These alarming statistics related to the wealth and financial security of African Americans caused our organization to develop a financial literacy initiative directed primarily to the African American population. The goal of the initiative, through financial literacy training, is to change financial behaviors, mindset, and outcomes in the acquisition and utilization of financial resources. Statistically, individuals and households with low levels of financial literacy are at greater risk of financial mismanagement than the general population.

The approach of SFE&PD's initiative is to reach African Americans at early adulthood. Specifically, the initiative was designed to infuse financial literacy education into the college experience of students at historically black colleges and universities. The students who attend these educational institutions are more likely to be first-generation college students who arrive from households where personal financial management concepts are not taught or discussed. Moreover, these institutions are located in states where the financial literacy level of the population is low or lowest. The Financial Literacy Index developed by the Financial Literacy Center using data from the 2009 National Financial Capability Study, a project of the Financial Industry Regulatory Authority Investor Education Foundation, indicates the level of financial literacy by state (see Figure 7.1).

SFE&PD's initiative is being implemented on campuses of historically black colleges and universities located in these states, but the program also reaches African American students from other states. The initiative includes a series of financial literacy seminars and workshops, called Mind Over Money Skills, on the campuses of 70 historically black colleges and universities located throughout the United States. The seminars and workshops expose young adults to the fundamentals, strategies, and tools of personal financial management that

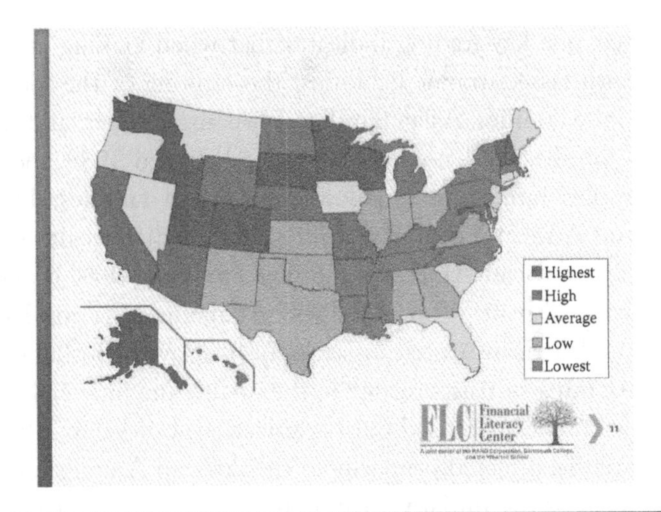

Figure 7.1 Financial Literacy Index by State. (From http://www.finra.org/newsroom/speeches/052913-remarks-about-national-financial-capability-study, retrieved April 15, 2015.)

they may not get elsewhere, or may learn too late. The general thrust of the series is to direct students to use financial resources to ensure financial sustainability and create wealth. Figure 7.2 shows the location of the colleges and universities that participate in our initiative.

It is expected that these students will share the financial knowledge gained with their families and friends and raise the financial literacy of the community at large. Over the years, after attending our financial literacy training, our students have shown improvement in their financial behavior and do share information with their family members and friends. We have received tremendously positive feedback from the students, educators, and community groups for whom we have conducted financial training, and we continue to implement their suggestions into our training. The evaluation form distributed to students after each training session asks the question: "Will you share the financial knowledge gained from the training with your parents, relatives, and friends?" The chart shown in Figure 7.3, which was developed from our most recent financial literacy training sessions held during the 2013–2014 academic school year, indicates that an overwhelming majority (90%) stated that they would share the information and financial knowledge gained with their families, relatives, and friends. This is an indication of our initiative's impact to enhance the financial health of the African American population.

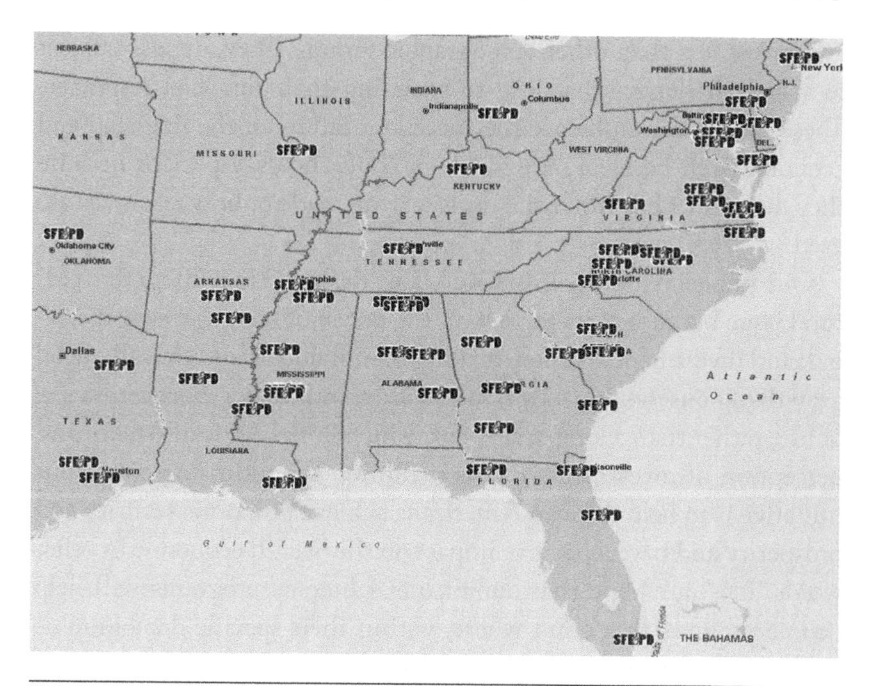

Figure 7.2 Location of colleges and universities participating in the Society for Financial Education and Professional Development programs.

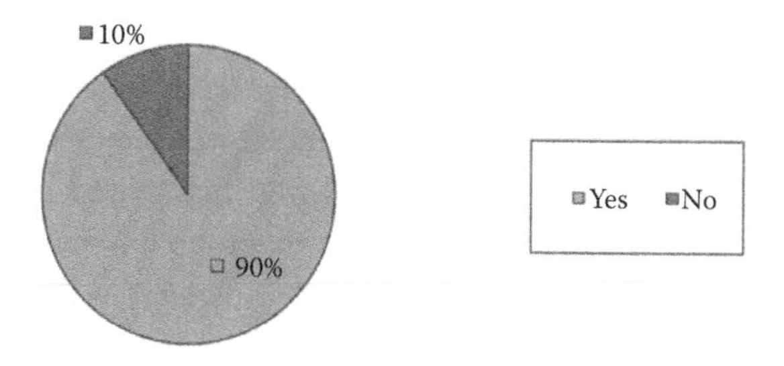

Figure 7.3 Chart showing that an overwhelming majority (90%) of participants state they would share the information and financial knowledge gained with their family members and friends.

The propensity to spend is often shaped by one's experiences and culture. The individual who grows up in a household that casually spends money without any particular goals in mind will do the same when he/she becomes an adult. In fact, this pattern of spending may continue from one generation to the next. There are data indicating that the African American population is culturally programmed to spend a greater portion of their income on consumer items, and save and invest less than other demographic groups. They are also subject to social challenges that lead to spending their financial resources. These factors can influence the success or failure of the financial and economic well-being of individuals and households, especially because they may affect how financial decisions are made or how resources are used.

Our financial literacy initiative is also designed to address this cultural issue because of its impact on the behavior of consumption, saving, and investment decision-making of individuals and thus financial growth of households. In fact, such culture indicates a short-term view of life, rather than a long-term view, which limits African Americans' perception of investments for their future. The ultimate goal of our initiative is to help African Americans achieve economic security and prosperity and have a positive impact on the overall economy. In other words, it is our hope that our financial literacy programs will help people progress to a point where, within their specific demographic group or culture, they can overcome challenges arising from cultural diversity so that they can enjoy a prosperous life in terms of personal and family financial well-being. The curriculum developed for our initiative includes provisions for:

1. Acknowledgment of the culture of the group relative to financial and economic literacy training;
2. Identification of resources available and that a particular group can use to enhance their financial well-being or are not aware of to enhance their financial well-being; and
3. Willingness or unwillingness of the group to seek help to enhance their financial and economic financial well-being.

Such provisions are included because financial education has to be in tune with the particular culture for which the training is presented. These provisions are important to our efforts to increase the effectiveness of our financial literacy initiative and financial decision-making. Lastly, we do consider that the values of African Americans may be in conflict with the broader society's economic values. However, through the acquisition of new financial knowledge, information, and strategies, this can be changed.

SFE&PD's financial literacy initiative also includes provisions for identification and understanding of one's values. Knowing one's values affects financial decision-making. If a person values owning a place to live, he or she will limit spending and save to purchase a home. If one values education, he or she will use his or her money to acquire education. If a person values keeping up with the "Joneses," he or she will spend a lot of money and save very little and probably live a life as a renter as opposed to a homeowner.

The initiative emphasizes the following financial core values indicated in Figure 7.4. Practical financial values help you build a life of financial growth, security, and contentment. Moreover, financial values enable individuals to bounce back from financial setbacks and temporary dislocations, including unemployment. In fact, financial values enforce financial self-control, delay gratification, and reject of impulsive buying. Each of these values can result in financial stability and reduce financial stress and/or anxiety.

Over the past 50 years, the average wealth in America has grown for all racial groups. However, this growth has not been equal for all groups. Figure 7.5 indicates the median wealth of three groups in America during 2007–2013.

There are several reasons why wealth has not grown equally. Many will immediately assume that income disparities are the causes of wealth disparities. This is not necessarily the case, as there are individuals who have high income but limited wealth. This is caused by poor credit management that limits homeownership, and unlimited spending that leads to limited savings and available funds to take

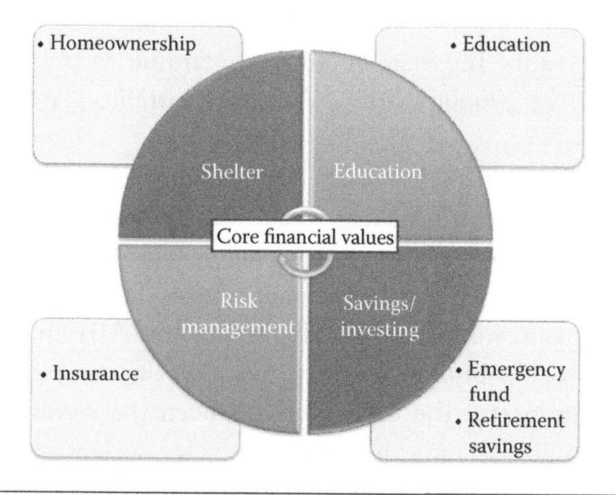

Figure 7.4 Core financial values.

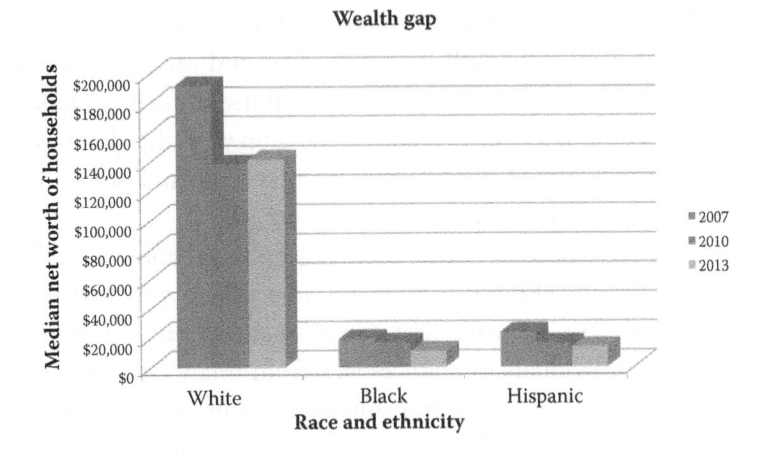

Figure 7.5 Median wealth of different racial groups in America in 2007–2013. (From Pew Research Center, available at http://www.pewresearch.org/fact-tank/2014/12/12/racial-wealth-gaps-great-recession/, retrieved April 15, 2015.)

advantage of investment opportunities. In fact, only 44% of African Americans own a home, which is the basis of wealth for many in the country. Much of this is the result of limited financial knowledge of the components of personal finance such as budgeting, establishing financial goals, cash flow management, savings, investments, credit and debt management, and risk management—acquiring appropriate insurance products to protect one's current income or assets.

The debt load of African Americans also limits their ability to create wealth. The high accumulation of student loans and the acquisition of predatory loans limit individuals and families in their ability to generate wealth. The share of American families with student loan debt was highest among African American families compared with other groups.

The high level of loans and their costs drains financial resources from households, which reduces funds available for accumulation of funds for a down payment on a home, emergencies, savings, investing, and participation in employer retirement plans.

There are many instances where higher-income African Americans do not have high debt loads, but have lower wealth compared to their White counterparts. The difference has been the result of investments made in conservative investment vehicles, such as certificates of deposit, saving bonds, and savings. In contrast, their White counterparts tend to make investments that are associated with more risk, such as stocks and bonds, which over the years have had a higher return than conservative investments.

Each of these scenarios is driven by limited financial knowledge. Because of these factors, our initiative focuses on providing face-to-face financial literacy training for African American students. This method of training affords our organization the opportunity to address the financial issues and concerns of the African American community immediately and provide instructions on how to resolve such issues, maximize financial resources, avoid the misuse of available financial resources, and discuss consumer protection.

The financial literacy seminars and workshops are designed to:

- Relate financial education to students' realities,
- Create new experiences and financial opportunities for students,
- Provide financial knowledge to avoid mistakes that impact students during and after school,
- Makes sense of financial concepts, and
- Provide "need to know" versus "good to know" financial information.

The financial literacy initiative introduces first-generation college students to critical personal finance and economic issues and practical financial management strategies and skills they can use in everyday life. Our approach incorporates the following elements:

- Practical application. Our seminars and workshops are not theoretical; we focus on germane information that adults encounter throughout their lifetimes.
- Targeted audience. Our seminars focus on young African American adults, enabling us to influence the culture of financial decision-making for a specific demographic. We study the latest literature and statistics so that our workshops hone in on African American personal finance habits.
- Interactive, lively format. In one example, students read aloud the terms of several credit cards. We then ask the group to determine which card they would select and why. Exercises

such as this engage the audience, demonstrate an understanding of the information being presented, and encourage reasoned decision-making.

- Highly skilled instructors. The individuals who lead our sessions are working practitioners who hold graduate and post-graduate degrees and certifications in finance, investments, and other related fields.

Each seminar/workshop focuses on personal finance topics and is culturally relevant. Our goal is to enable students to make informed decisions regarding spending, saving, investing, risk management, and the use of credit. Our most popular sessions are geared with the ages of young adults in mind:

- *Credit Management* is designed for incoming freshmen. Students learn the responsibilities associated with establishing credit and how to make informed decisions about spending. This workshop includes the latest information about credit cards, student loans, and credit scores. Importantly, it conveys the impact of students' current credit decisions on future employment and wealth accumulation.
- *Personal Money Management* is geared toward upper-class students who will soon enter the workforce. Up to 2 hours long, this session focuses on how to leverage a college degree for the benefit of personal financial growth. Topics range from issues students will face in the upcoming months (budgeting, saving, investing, insurance) to those they must plan for in the coming years (homeownership, retirement planning, estate planning).

The result is an information-filled, high-energy workshop that college student participants say they rely upon years after they attend. These seminars are presented on campus and designed to integrate seamlessly into each institution's curriculum or extracurricular program schedule. Each seminar is offered during the academic school year and typically presented in a classroom or lecture hall. All educational materials, including handouts, are provided. Additionally, provisions are made for question-and-answer periods and one-on-one problem-solving consultations with students.

Financial Literacy Experiences

During the implementation of our initiative, we have found that many individuals make financial decisions based on emotions rather than sound financial knowledge. We often state that financial decisions should not be based on emotions, chance, or luck. Each financial decision must be made based on sound information, including a complete calculation of the cost of the item or service purchased or opportunity costs. We often tell our audiences to remove their emotions and "run the numbers," which means to determine the true cost of items or a business transaction before a purchase is made to ensure that your money is efficiently used or maximized.

Outcomes from Initiatives and Experiences

Almost every day, SFE&PD receives personal feedback from individuals and students who have attended our training sessions regarding the effectiveness of our initiative and how our financial literacy workshops are making a difference in their lives. For example, SFE&PD recently received a call from a young man who attended one of our seminars while in college and he told us he earned $40,000 from his investment in his company's 401(k) plan last year. An elderly mother who attended a seminar at a church in Atlanta, Georgia, in 1993 recognized SFE&PD staff at the same church just last year and came up to thank us for changing the life of her family.

In addition, testimonials from students and professors who attend our seminars let us know that we are successful in helping them make better financial choices by teaching them fundamental financial principles. A professor from Xavier University said, "In these tough economic times, presentations like these become even more important for our students. They need to be made aware of money management issues so that they make all the right decisions as they go through their student life and enter careers." Student feedback from the workshops let us know we are reaching college students and providing the appropriate financial knowledge they need to help them plan for their future. A student from Howard University's Dental School summed it up best, "Growing up, I was never a huge spender. The day I started dental school was when my world changed. Until Mr. Daniels' (presentation), I had no clue and was financially illiterate. Now, after I had an overview of proper financing, I am ready to take on the world!!! I am thinking long term now thanks to Mr. Daniels!" An undergraduate at Bowie State University told us that the seminar she attended was "informative and engaging.... I believe that all college students need financial training and I know I need to start my financial planning for the future now!"

Evaluation Measures

After each of our sessions, students complete a survey as feedback on the training session. We prepare an analysis of evaluation responses to determine the effectiveness of each seminar. Program evaluation of the following seminars provided the following feedback to support the effectiveness of the program.

- When students who participated in our personal money management seminar were asked the question—"Will the information presented help you better manage your finances?"—96% of the participants responded "Yes."
- When students who participated in our credit management seminar were asked the question—"Will the information presented help you better manage your credit?"—95% of the participants responded "Yes."
- Based on student evaluation forms submitted by upper-class college students, 99% stated that they would make investments in the future.

In addition, our program has been recognized on a national level for its educational contributions. SFE&PD was extremely pleased to received the prestigious Federal Deposit Insurance Corporation's Chairman Award for Excellence and Innovation in Financial Education in 2009. The U.S. Department of the Treasury also presented us with a certificate of recognition, which states that our financial education seminars meet the criteria for effective financial education programs identified by the Department's Office of Financial Education.

Future Trends in Financial Education

The future trend in financial education is research based and presented in several formats such as financial games, online instructions, face-to-face instructions, and peer-to-peer and group sessions. However, there needs to be more development of intervention tools such as financial apps, wizards, use of social media, and public service announcements to help individuals make sound financial decisions. The products developed should address the practical application of personal financial management concepts, techniques, and strategies. The use of nonprofit organizations in financial education is very significant because such organizations are able to present unbiased financial education to individuals and groups. In many instances, individuals are more accepting of the information provided if they feel the financial education presented is not associated with the sale of a product or service. The challenge of nonprofit organizations in youth financial education is acquisition of funds to develop, implement, and sustain its programs.

Gaps in Financial Literacy Education

The gaps in financial literacy education are at the primary, secondary, and postsecondary education levels. Because financial education is not a part of the Common Core Education requirements, very little time is available to institute financial literacy programs or other programs that would provide important financial knowledge and skills to students.

SFE&PD strongly believes that the opportunity to build foundational financial education should begin at the early grade levels of school when the fundamentals of reading, writing, and mathematics are taught nationwide. For example, a reading lesson might include financial terms and scenarios that would enhance a student's knowledge of personal finance or economic principles. This could be effective even if mandatory stand-alone personal finance courses are not taught in school systems. Additionally, training teachers to teach financial education is another gap in financial literacy education that must be addressed. Typically, most teachers are not prepared to teach personal finance and economics or have a limited mastery of the subject matter. Those teachers may have had fewer than one or two undergraduate courses in these subjects while pursuing a bachelor's degree in education, which makes it difficult for them to teach financial education to their students. Provisions should be made to integrate personal finance into bachelor degree programs to remedy this problem or develop special financial education programs for existing teachers that include incentives for participation.

Although many institutions of higher learning are increasingly aware of the need to provide financial education to students because of the financial pressures and high educational costs that college students are experiencing today, most institutions only direct their efforts to providing education in students' fields of study. However, a few colleges and universities that are ahead of the curve are offering a personal financial management course as an elective for the entire student body. The provision of financial education at the postsecondary level can increase an institution's student retention and graduation rate and can have a lasting positive effect on college students. This is why building public support and awareness about the need to implement financial education at colleges and universities is so important. There are also opportunities to provide financial education at the postsecondary level, which could take place at "teachable moments" and required assemblies with targeted subjects, such as credit card and student managements, budgeting, banking, automobile purchase, cell phone use, retirement planning, and other important personal finance subjects.

Managing money is a large part of each individual's life. Therefore, it is essential to acquire financial knowledge and understand the practical application of personal money management concepts. Individuals who understand and use these financial concepts and strategies do well in managing, growing, and protecting their long-term financial resources. This is why we conduct financial training for college students. SFE&PD believes that it is essential to reach as many young people as possible at the early stages of life to assist them on a path to economic independence and stability. The continuous implementation of SFE&PD's financial literacy training initiative can empower generations of young people to have more successful lives with financial stability and wealth growth.

References

Boshara, R., and Emmons, W. (2013). After the Fall: Rebuilding Family Balance Sheets, Rebuilding the Economy. Federal Reserve Bank of St. Louis. Retrieved from https://www.stlouisfed.org/~/media/Files/PDFs/DWTF/After-the-Fall-5-23-13.pdf.

Boshara, R., Emmons, W. R., and Noeth, B. J. (2015). American Incomes: Demographics of Wealth. Federal Reserve Bank of St. Louis. Retrieved from https://www.stlouisfed.org/~/media/Files/PDFs/HFS/essays/HFS-Essay-1-2015-Race-Ethnicity-and-Wealth.pdf.

Ryan, B., Osaki, Y., Burhouse, S., Chapman, D., Critchfield, T., Goodstein, R., Samolyk, K., Harris, A., Reynolds, L., Gregorie, L., Bachman, M., Gill, P., Glenwick, M., Grazal, J., Johnson, P., Spanburg, D., Villar, J., and Zeidler, K. (2009). FDIC's National Survey of Unbanked and Underbanked Households. Federal Deposit Insurance Corporation. Retrieved from https://www.fdic.gov/householdsurvey/2009/executive_summary.pdf.

Schmeiser, M. D., Buchholz, D. E., Brown, A. M., Gross, M. B., Larrimore, J. H., Merry, E. A., and Thomas, L. M. (2013). Report on the Economic Well-Being of U.S. Households in 2013. Federal Reserve Board. Retrieved from http://www.federalreserve.gov/econresdata/2014-economic-well-being-of-us-households-in-2013-preface.htm.

8

Financial Literacy Education at Harrisburg University of Science and Technology

A Case Study

JAY LIEBOWITZ

Contents

Financial literacy should be an important part of an individual's education, especially in high school, college, and beyond. At Harrisburg University of Science and Technology (HU), our students are all Science, Technology, Engineering, and Mathematics (STEM) majors, whereby many of our undergraduates are first-time college seekers in their families. Through the endowment of Mr. Alex DiSanto, a prominent developer in Central Pennsylvania, the focus starting in August 2014 was to instill financial literacy and other business education to our students, as well as to Harrisburg high school students and the local community. The DiSanto Chair was established to accomplish this mission.

This chapter will highlight the various approaches that we have used to initiate financial literacy education at HU and the surrounding area. It may serve as a model for other similar institutions.

Targeting Our HU College Students

Because our students are STEM majors, we wanted to start to infuse financial literacy early in their college career. HigherOne's and Everfi's "Money Matters on College Campus" annual report shows a statistical significance between introducing financial literacy early in one's education (i.e., high school/college) and increased responsibility in terms of a college student and young adult being financially responsible (such as paying back their college loans). We are including financial literacy throughout all 4 years of the undergraduate education. In the freshman year, our students are introduced to National Endowment for Financial Education's (NEFE) "Cash Course" (cashcourse.org), an online course that helps students better understand personal finance. In the sophomore year, our students are introduced to more higher-level concepts through our general education courses, such as the Organizing Mind. We introduce the concepts of compound interest, capital budgeting techniques, creating a short-term, mid-term, and long-term personal financial plan, Roth IRAs (Individual Retirement Accounts), various financing options for starting a business, and other financial literacy and microeconomics considerations. At the junior and senior years, the students are introduced to other concepts through required seminars that cover a variety of topics to help the students achieve success in their careers. We also bring in speakers into the classroom—such as Gene Natali, an investment banker who wrote *The Missing Semester*, which features financial literacy concepts; and Doug Hassenbein, the assistant investor education coordinator from the Pennsylvania Department of Banking and Securities.

To complement this formal financial literacy education in the classroom, the DiSanto Chair Distinguished Lecture Series in Financial Literacy was established, whereby every month we invite speakers from the outside to expound on various topics related to financial literacy. The speaker series is free and open to everyone, including the local community, high school students, and beyond. The fall 2014 inaugural Speaker Series focused on a macroview of financial literacy, whereby we invited senior-level executives from HigherOne, National Association of Securities Dealers Automated Quotations (NASDAQ), and Financial Industry Regulatory Authority (FINRA) to come over and discuss their perspectives on the economy, markets, bitcoins, insider trading, college awareness programs for financial literacy, and other related topics. The Spring 2015 Speaker Series focused on a microview, whereby we had panels and lectures on banking regulations (bank presidents), financial fraud and creative accounting techniques (chief of the Audit Division for Pennsylvania's Department of Enforcement and Investigations), and entrepreneurship advice on starting your own high-technology business (DC Founder's Institute) (http://www.harrisburgu.edu/academics/financial literacyspeakerseries.php).

During such events, we typically try to fill the HU auditorium, which can accommodate up to about 130 attendees. The talks are also webcasted and preserved in our video archive for easy access after the event.

We are also involving HU students in a number of financial literacy-related competitions. For example, the Chartered Financial Analyst (CFA) Society in Pittsburgh has a financial plan competition, and HU students have entered their personal financial plans as part of the competition. An entry submitted by one of our students was chosen as one of the top 10 entries out of more than 120 personal financial plans sent in by participants. Also, as peer-to-peer mentoring seems to be a preferred approach for financial literacy education success (according to the research), we introduced a peer-to-peer mentoring contest to develop the most appropriate model for HU to use. The best approach would be selected, including educating the student mentors, and it is our hope that we would be able to implement this model at HU with various student mentor–mentee pairs.

Targeting High School Students in Harrisburg

As the literature points to the value of introducing students to financial literacy education in secondary schools, we wanted to reach out to inner-city students in the Harrisburg School District to help them become more financially literate. For starters, we met with their school board and introduced the "What Does Financial Literacy Mean to Me?" contest for Grades 7–12 in the Harrisburg School District. We invited students to submit a video (up to 2 minutes), essay, poem, or short story relating to the theme of the contest. We received 21 entries, and three winners were selected and awarded cash prizes ($500 for the first place winner; $250 for second place; $125 for third place). All participants received a certificate of appreciation. The first place winner's submission can be viewed at https://www.youtube.com/watch?v=3kVPMkCGars&feature=youtu.be.

We are also exploring relationships with the local Jump$tart and Boys & Girls Club in the Harrisburg area to further introduce financial literacy into their programs.

Targeting the Local Harrisburg Community

HU also wants to extend financial literacy learning opportunities to those in the local community—businesses, parents, teachers, government officials, etc. The DiSanto Chair Speaker Series is open and free to everyone. We have also established the Analytics Applications Summit at HU, which is a free 1-day event that features chief analytics officers and other leading individuals who hold discussions on various topics (e.g., financial analytics, social media analytics, healthcare analytics, big data analytics, marketing analytics, and other types of analytics relating to financial literacy education). More information is available at this link: http://www.harrisburgu.edu/academics/undergrad/analytics/analyticssummit.php.

The Analytics Summit has about 120 attendees, mostly from various organizations in the area, as well as some of our HU Analytics students.

We also have increased marketing coverage for financial literacy education in the local community, such as doing a segment on financial literacy on the local Fox television affiliate (see http://fox43.com/2014/12/16/teaching-financial-literacy-at-a-young-age/).

Possible Research Projects in Financial Literacy at HU

One research project that has been proposed at HU is "A Virtusphere Approach to 3-D Game for Investing for College Students." The primary goals of the project include:

- To promote college success and financial literacy education via a three-dimensional (3-D) virtual reality immersive "investing game" environment, as well as in walking through the trading floors

- To complement and extend our focus on financial literary education, via the newly endowed Chair and the DiSanto Chair Distinguished Speaker Series in Financial Literacy, to promote financial capability and asset building

The target audience is primarily Harrisburg University college students, and other high school and college students in the area. Certainly, those in the local community are also welcome to experience the Virtusphere and learn about investing. Our plan is to work with the Center for Advanced Entertainment and Learning Technologies, led by Charles Palmer, at Harrisburg University. The students and faculty associated with the Center are expected to be involved in the development of this application.

HU caters to college students, mainly underrepresented groups, who are interested in science and technology. To prepare them for technical entrepreneurship or just handling personal finance, they need to be better educated on financial markets and investing. Toward this goal, the proposed project would involve using the Virtusphere to allow students to experience an immersive environment in which we will develop an "investing game" and also allow them to virtually experience the trading floors.

We are one of only four universities worldwide to have a Virtusphere. Using the Virtusphere, we are able to leverage learning theory and practices with an innovative learning technology that transforms the learner's world into a virtual reality that brings learning to life. Our faculty can design and author interactive simulations for the Virtusphere, using the newest game engine software, and recreate 3-D models of any environment you need to explore.

According to Professor Palmer, the immediately obvious feature of the device is a 10-foot, hollow ball of ABS plastic. The user enters the ball through a hatch with additional hardware that includes wireless head-mounted display goggles. In Virtusphere, users can safely traverse a real or imagined 3-D world in life-size or miniature. Sensors monitor the speed and direction of your steps as well as which direction you look. The image you see on the head-mounted display shows a changing view wherever you walk and look. A light plastic controller can also be carried for sending signals and targeting. Other people can view the scene on monitors outside the Virtusphere; the experience can be recorded for future use as a training and evaluation tool.

This 12-month project would have the following deliverables: the actual investing game within the immersive Virtusphere environment, final report, and presentation.

Results would be evaluated by pre- and post-tests in working with HU college students in determining their financial knowledge gained from use of the Virtusphere and investing game. We would also have them develop a 1-year, 3-year, and 5-year financial plan and compare their budgeted plans to their actual plans during their college career and upon graduation, through surveys and follow-up interviews. Funding for this research project is in its exploratory stage.

Other initiatives at HU involve encouraging the Commonwealth of Pennsylvania to include a personal financial literacy course as a requirement for high school graduation. As of February 2015, only 17 states in the United States have this requirement. Even though April is "Financial Literacy Month" in Pennsylvania, as well as in many other states, the Commonwealth should strongly encourage making this a requirement for high school graduation. In recent years, there has been legislation in Pennsylvania to make this a reality; unfortunately, the bill has not been passed to date. At HU, we are fortunate to have two new vice presidents who were very senior executives in (Pennsylvania) Governor Corbett's administration. With their connections, we hope we can make some inroads into this issue.

Challenges and Opportunities for Financial Literacy Education at HU and Elsewhere

We often hear that some of the key issues in financial literacy education include teacher training and associated budgets, integration of the material, and employers. In secondary schools and higher education institutions, teachers and professors of economics, business, management, or allied fields may often be called upon to teach financial literacy in these schools. However, many faculty members may have to be retrained and reeducated on how best to cover financial literacy as part of the curriculum. In relation to the curriculum, another challenge is how to best integrate financial literacy throughout the curriculum. For example, at HU, we are integrating financial literacy through the microeconomics material, as well as into our "Success" seminars for all undergraduates over the 4-years of their college education. We must also look for ways to further integrate this material into the other major courses, including information systems, biotechnology, business entrepreneurship, and general education. Finally, another major challenge is how to engage the employers into providing the added insights and job/internship opportunities for those who are interested in weaving financial literacy into their respective majors. The employers are an integral part of our financial literacy education ecosystem.

To address some of these challenges, various approaches can be used. First, the Student Financial Aid Office at the school or university can play a major role in further exposing the students to financial literacy and resulting loan obligations. Their work will complement the faculty's formal discussions in the classroom. Second, mentoring programs can be established to promote both peer-to-peer mentoring in financial literacy and student–employer/faculty mentoring as well. By engaging the employers in the lives of students, the financial literacy education of the students will be enhanced through real-world situations and practical advice throughout the student's college career. Third, if more states pass legislation to include a personal finance course as a requirement for high school graduation, then hopefully the appropriate resources (both money and labor) will be allocated in State budgets to help achieve this goal. Lastly, reaching out to the parents and families of the students and involving them in financial literacy education will greatly assist in leveraging the concepts discussed in school and applying them at home and at work.

Summary

Various colleges and universities, such as Allegheny College, University of Pittsburgh, and Ohio State University (as highlighted in this book), are already either requiring or strongly encouraging their students to learn about personal finance and financial literacy education in general. At HU, we are using an integrative approach to weave financial literacy education throughout the 4-year education of our college students. We are also trying to involve our graduate students in this area through the various Speaker Series, symposia, and competitions that are open to them as well as our undergraduates.

It is still too early to predict success at this point, as we are only in our first year of our financial literacy efforts. Anecdotally, we are told by students that we have opened their eyes to important topics on this matter and they, in turn, have raised some of these issues through discussions with their families and loved ones. We will be doing a more formative analysis and assessment in the near future in order to measure the impact of these learning opportunities through a longitudinal study. We are optimistic that we will be successful, and feel satisfied that we have contributed to the financial security and responsibility of our students and their future growth.

9
Don't Just Survive—Thrive!

PATRICIA W. COLLINS

Contents

History

Sam Houston State University (SHSU) is a small regional state college in east Texas with a student population of more than 19,000. Many of these students use federal student loans to pay for their college education. While working with student inquiries regarding the timing of their financial aid refund, the director of the Bearkat OneCard office noticed that students did not seem to be very informed about financial aid or banking processes. A qualitative student survey was processed indicating that students recognized the need for a financial literacy component in their education. A focus group of students confirmed this need and indicated they would be willing to pay for financial literacy services. Through a collaborative partnership with Student Services, Academic Affairs, and Enrollment Management and support from the Texas State University System, the SHSU Student Money Management Center (SMMC) opened its doors in October 2008 to the students of SHSU. The mission of the SMMC is to empower the students of SHSU with the unbiased education and tools they need to achieve financial independence.

Financial Literacy Outreach and Programming

The SMMC offers financial literacy outreach to our students in the form of e-learning modules, podcasts, workshops, collaborative classroom presentations, and one-on-one professional and peer coaching. The topics covered include, but are not limited to, the following: budgeting, types of credit, understanding credit scores and credit reports, personal net worth, debt-to-income ratios, identity theft protection and cyber security, saving and investing, retirement planning, insurance, financial goal setting, starting your own business, buying your first house, and financial aid.

Staff Training

To ensure that these topics are properly addressed, each professional counselor on staff is certified through both Association for Financial Counseling and Planning Education (AFCPE) and/or the Institute for Financial Literacy. Peer coaches are likewise certified by the

Institute for Financial Literacy and also undergo an extensive training program provided by the professional staff members.

Collaboration—External

Collaboration is the cornerstone of the Center and has been critical to the success of the SMMC. The SMMC was one of the first three collegiate money management centers established in the State of Texas. Texas Tech University, the University of North Texas, and the SMMC shared research and programming ideas. There is no doubt that the strength of collaboration was extremely beneficial. In order to strengthen this collaborative effort, the SMMC has shared its research and programming with other universities in Texas as well as universities throughout the United States. In 2013, the SMMC hosted the inaugural conference for the Texas Association of Collegiate Financial Education Professionals. Each year, a symposium is held so that collegiate professionals can stay informed about the changes in regulations, resources, and programming concerning personal financial management.

Internal Collaboration

Internal collaboration has firmly entrenched the curriculum of the SMMC into the academic and student services education of the student body. Internal collaboration develops a presence on campus and garners a network of referrals. A few of these are mentioned in the following sections to show the various types of service the SMMC can provide to departments and students that bring meaning and definition to the Center.

Internal Collaboration—College of Education

The College of Education has been awarded a TRIO Project CONNECT grant from the Department of Education. Financial literacy is one of the required components of the grant, and the SMMC works with Project CONNECT to provide budgeting and planning skills for their students to encourage retention at the university. The SMMC also collaborates with the College of Education to train their mathematics student teachers in the basic financial skill sets required by the Texas Essential Knowledge and Skills. The state of Texas recently mandated that budgeting, saving, credit, and methods for paying for college should be added to the mathematics K–12 curriculum. The SMMC developed a 3-hour curriculum, along with pre- and post-tests, and certifies that the students have been trained in financial literacy. In addition, the College of Education and the College of Criminal Justice invite the SMMC into their classrooms to instruct juniors and seniors about federal loan repayment programs such as Income Driven Repayment Plans and Public Service Loan Forgiveness. The SMMC also recently filmed a video that has been posted to their BlackBoard and graduate school's website. The video explains different methodologies that make graduate school affordable such as graduate assistantships, scholarships, the TEACH (Teacher Education Assistance for College and Higher Education) Grant, and Public Service Loan Forgiveness.

Internal Collaboration—Student Advising and Mentoring Center

In the 2010–2011 academic year, the Student Advising and Mentoring (SAM) Center piloted a program called SH Elite. The program's primary goal is to increase the retention rate and graduate men of color. The program has been very successful. Initially, the program was for freshmen only. In time, however, the program grew. Upperclassmen stayed in the program to serve as mentors, but the SAM Center wanted to offer services to these upperclassmen as well. The SMMC worked to develop additional financial literacy programming tailored for the upperclassmen so they would not participate annually in programming previously provided. The SMMC presents Financial Aid 101 to freshmen, Personal Goal Setting and Budgeting to sophomores, Buying Your Own Home to juniors, and Investments to seniors. The personal connections developed with students in this programming have led to personal coaching sessions where the students have learned to create additional revenue streams by completing scholarship applications and utilizing the work study program effectively. In addition, many of these students have set goals to reduce the amount of student loans they utilize over their college career. Some students limit their borrowing goals to using only Title IV subsidized loans.

Internal Collaboration—Student Food Pantry

The SMMC publishes a free food calendar to help students recognize a way to reduce their personal expenses as well as fight food insecurity among our student population. Copies of these calendars were placed on billboards throughout the campus, as well as offices across campus including the Student Food Pantry. A Student Food Pantry board member approached the SMMC about partnering to bring the Houston Food Bank to the SHSU campus. The SMMC created a "Feed the Growl" marketing campaign and worked with campus representatives and food bank personnel to develop logistics. More than 300 people were served that day, and 34 students completed applications for food stamps.

Internal Collaboration—First Year Experience

First Year Experience champions partners with freshman seminar classes and various services on campus. Financial literacy is offered as a topic taught by the SMMC. A variety of topics are offered for professors to select. The most popular are student loans, budgeting basics, and 12 tips to living rich. Classes are taught by either the Kat2Kat peer coaches or staff members.

Internal Collaboration—College of Fine Arts and Mass Communication

The College of Fine Arts and Mass Communication collaborates with the SMMC on a variety of levels. Their mass communication students often interview SMMC personnel for radio and television programming. In addition, students are assigned to film and write newspaper articles or reviews of SMMC workshops. Their students may also earn extra credit by participating as emcees or photographers at SMMC events.

Internal Collaboration—College of Business Administration

The SMMC has embarked on a new endeavor with the College of Business Administration (COBA). The SMMC is calling this new endeavor Train the Trainer. COBA students know it as Plan for $uccess. The faculty advisors and students of Alpha Kappa Psi and the

National Association of Black Accountants have agreed to participate in a community service project to provide budgeting workshops and consultations for clients of the local United Way agency. The project will be implemented in the upcoming fall semester. The SMMC will develop the curriculum and provide training for the students to speak at the budgeting workshops and provide personal consultations. The opportunity will not only provide financial training to the SHSU students and citizens of the community, but also will provide the students the chance to develop their knowledge, enhance their public speaking skills, and gain experience in coaching on personal financial topics.

Internal Collaboration—School of Nursing

Through interviews with potential students entering the nursing program, the SMMC identified that these students were at risk for losing eligibility for their financial aid. Many times, students were advised to retake classes to increase their grade point average (GPA). This decreased the students' completion rate required for Title IV federal aid. In addition, many students transferred into the nursing program from other majors or other schools. Some of these students were edging closer to the maximum hour limit. The SMMC met with advisors in the nursing program to educate them about these issues. Students with these issues are identified early and referred to the SMMC to begin identifying different income streams. The SMMC took on the task of mining scholarships for students in the school of nursing and provides an annual reminder when invited to provide a financial literacy workshop to the Student Nurse Association.

Internal Collaboration—Office of Student Financial Aid and Scholarships

The SMMC relies heavily on the knowledge base of federal and state financial aid. This is a strong skill set within the Center. A financial coach for college students must have a full deck of cards to extract for unique circumstances. Waivers for foster students, veterans' benefits, disability benefits, and scholarship search engines are just a few of the wild cards that can be applied in unique situations as they are recognized. It is imperative for SMMC personnel to have routine training and updates regarding different types of financial aid.

At SHSU, the SMMC collaborates heavily with the financial aid office. The SMMC works with financial aid and sits in on student appeal hearings. The SMMC also offers budgeting services for students who file an appeal for additional grant funding with financial aid. In return, the personnel in financial aid routinely refer students to the SMMC if they identify that students are struggling to make ends meet or inappropriately using their financial aid refund. The SMMC also refers students to financial aid if it appears there may be different types of funding for which the student may have eligibility.

Financial Literacy Week

In 2009, the SMMC premiered a solid week of programming called Financial Literacy Week. Internal collaborators such as the Office of Financial Aid and Scholarships, the Health and Counseling Center, Career Services, Veterans Resources, Student Legal and Mediation Services, and Student Leadership Initiatives were invited to make presentations such as Financial Aid 101, Love and Money, Healthy Meals on a Dime, Analyzing Your Job Offer and Benefits Package, The Law and Your Wallet, Having Fun on a Dime, and Pinterest Ideas for Reducing Your Expenses. External partners were brought in to discuss the following topics: extreme couponing, financing your ride, insurance, identity theft, and investments. Student organizations are awarded prizes based on percentage level of attendance at the workshops. In addition, students are offered the opportunity to win individual scholarships. Almost 1000 students attend this event annually.

You Can Build It; But, Will They Come?

The SMMC had built strong, researched curriculum and developed an internal and external support network. In 2011, the SMMC was recognized as the "Outstanding Financial Counseling Center of the Year" by the AFCPE. Student speaker and coaching surveys indicated that students were learning about financial literacy components. The surveys also indicated that programming and instruction were strong and that the respondents would utilize the services of the SMMC in the future; however, the level of student participation at SMMC-sponsored workshops and personal coaching sessions was low and was not increasing.

Personal Approach to Student Programming; Increased Level of Voluntary Participation

In 2013, the SMMC implemented a more personal approach to student programming. Indoctrination of freshmen into the SMMC culture at student orientation is considered mission critical. Focused personal marketing identifying the strategic services offered by the SMMC is imperative to the Center's success. The SMMC does not want our graduates to live from paycheck to paycheck. We want them to thrive! Marketing and programming for new student orientation was designed to convey this overarching message. New student orientation was the springboard for the 2013–2014 academic year of SMMC student success. On day 1 of the orientation, a full-time staff member attended the parent ice cream social and at least one full-time staff member and two peer counselors attended Passport to Sam—a carnival-like family introduction to student services and organizations. On day 2, a full-time staff member made a presentation at a voluntary orientation breakout session. A brief explanation of the services offered was given and specific, relevant budget and expense saving measures were discussed with parents and students. Immediately, parents and students came to counseling sessions to discuss their personal budgets. In addition, the SMMC offered to help them process student loan entrance counseling. Throughout the summer, attendance at orientation breakout sessions and counseling sessions grew. Title changes were made to make financial counseling more personal. Counseling became coaching; counselors became coaches.

Workshop Topics—Relevant and Timely

The SMMC 2013 fall workshop programming was developed to make topics relevant and time appropriate for students. The first workshop that resonated with students (and employees of the state university) was After Graduation: Student Loan Repayment. Personal invitations were sent via e-mail to juniors and seniors offering ways to potentially eliminate or reduce their student loan debt. The program covered cost aversion by avoiding late filing fees for graduation, state rebates for graduating on time, income driven repayment plans, public service loan forgiveness, teacher loan forgiveness, perkins loan forgiveness, and Veterans information about federal student loans. Students requested copies of the presentation to assist with processing student loans and rebates; they attended personal coaching sessions and referred friends for coaching. The SMMC also worked with SHSU Human Resources personnel to offer information about Public Service Loan Forgiveness to new employees of the university. This increased referrals to the SMMC and personal coaching sessions. The number of personal consultations in academic year 2013–2014 increased by 88% over the previous year.

Review of Workshop Viability and Restructuring

Attendance levels at workshops had not increased in the previous academic year, so they were reviewed for potential elimination. The review indicated that workshops were not only mentoring tools, but they should also be the core of the SMMC campus-wide marketing strategy. An intentional and deliberate plan of action was put into place to develop the marketing strategy. A small student focus group determined that the timing of the workshops should be changed to early evening to gain more student attendance on campus. Workshop times were changed to Wednesday evenings at 6 P.M. and free pizza was offered—first come, first serve. In addition, a new scholarship program was offered based on an individual student's level of attendance at workshops. This was mirrored after the successful marketing of the Financial Literacy Week events. At the end of the semester, a drawing would be held for two $400 scholarships for students who had attended at least three semester workshops. An Excel spreadsheet was developed to generate projected release times for cross-media marketing before and after the event: targeted e-mails to friends of the SMMC, flyers, photos, scrapbooks, website, social media, digital billboards, student newspaper, tabling events, and university public relations.

Increased Level of Information Provided at New Student Orientation

In March 2013, planning for workshops held in academic year 2014–2015 began. Subjects were planned, and marketing was created and distributed at new student orientations. Many incoming freshmen attended the first session of the fall. It was held on the first day of the class in conjunction with Welcome Week so it received multiple types of publicity. In addition, the programming was targeted to incoming freshmen: So You're On Your Own; Now What? The 2014–2015 academic year rounded out the workshop calendar in May with Weddings on a Dime. Students attended workshops because they were productive: the topics were timely and relevant to their financial goals. In addition, they decreased their expenses by eating free pizza and potentially increased their income by winning a scholarship based on attendance! Workshop attendance increased more than 44%.

Increased Level of Referrals and Attendance at Workshops

The marketing success for workshops empirically increased the number of referrals as well as presentations, and personal coaching sessions requested by professors, student organizations, and individual students. Although the Center is student-focused, marketing has also culminated in coaching sessions for faculty, staff, and alumni as requested. (Service to these personnel is also viewed as a form of marketing because they generate student referrals.) In the past year, with a staff of two full-time employees and three peer coaches, the SMMC conducted 29 center-sponsored workshops, 67 classroom/organization presentations, 584 student coaching sessions, and 51 recruitment/outreach events.

Most Common Coaching Topics

Budgeting and goal setting is the most common topic of personal coaching. The SMMC works with students to embrace being a poor college student and learning that it is important to reduce expenses while they are in college so they don't borrow unnecessarily and graduate with a large negative net worth. Students understand percentage-based budgeting. They also comprehend goal setting topics such as (1) saving 10% of their salary for 2½ years after graduation results in saving 3 months of their net salary and (2) if they graduate at age 23 and save 10% of their salary, they will save more than 1 year of their net salary by 35 years of age. As students gain the skill sets for budgeting and goal setting, their interest in other financial literacy topics grow. For example, they will request follow-up consultations or workshops about savings and investments or how to buy a house.

Student loans are another popular topic of personal coaching. Students may have different goals in this regard: (1) requests for ways to increase or decrease loan amounts borrowed or (2) methods for budgeting student loan repayment. In many instances, personal coaching reveals that students may not need additional loans if they increase other income streams or reduce debt.

Surprising Coaching Revelations

Some students indicate they processed their federal aid, but were not aware that they had to pay back federal student loans. Multiple students have indicated that they knew they had used financial aid to help pay for college, but were unaware that they had processed student loans. Students do not remember processing loan entrance counseling. Some students indicate that perhaps their parents had processed the free application for federal student aid and also processed the student loans on their behalf.

Conclusion

The path to a successful SMMC is deliberate and intentional. Collaboration, marketing, and quality outreach and programming are critical elements. However, the most important aspect of a personal financial literacy program is the individual student. Students' financial goals and priorities are different. The staff and peer coaches must actively listen to students' financial needs and concerns such as food insecurity or ability to pay student loan debt. Then, the team must proactively research the issues, potential resolutions, and develop programming and methods to educate their students on these topics.

College graduates should not just live from paycheck to paycheck. They should thrive! They should be taught how to aggregate wealth, not debt. They should be taught how to develop income streams and reduce expenses. This empowers students to have a positive financial net worth and accomplish their financial goals.

10

Collegiate Financial Literacy

The Ohio State University Example

BRYAN ASHTON

Contents

As financial literacy increasingly becomes a hot topic across the country, higher education has not been exempt from the conversation about the need to better prepare students to face their financial future. This pressure to better prepare students for their financial lives is further fueled by the increase in student loan debt that students assume to attain a degree, the increase in student loan default among those who borrow for educational purposes, low financial knowledge levels (Trombitas, 2012), and the perceived societal increase in pressure to create a financial return on investment for a college degree. Additionally, there is an increasingly common understanding that individuals are entering into large financial decisions (going to college) without having the adequate information to make the decisions and understand the long-term implications of such decisions. As such, there has been a trend on campuses that shows an increased agreement among administrators around the need to address the shockingly low level of financial literacy in their student body (Trombitas, 2012).

As this issue continues to evolve, collegiate financial literacy programs have taken numerous approaches to attempting to reach students and design interventions for their student population. These approaches tend to include the distribution of information via online vehicles, presentations and workshops, one-on-one coaching/counseling, and at times curricular options. Just as the K–12 space is faced with the challenge of enrolling a large number of individuals in a stand-alone course, higher education has traditionally used one or two time interventions to engage students in financial information. This inability to sustain engagement has prompted topics of conversation about outcomes as programs attempt to define their impact and set learning outcomes related to the delivery of financial literacy information within the confines of student scheduling limitations.

Despite some of these challenging factors, there remains an increased pressure to provide real-world skills for students, which has led, and will continue to lead, to an increase in collegiate financial literacy initiatives. These initiatives have taken a variety of forms, with a variety of expected outcomes, including decreased student loan borrowing, decreased default rates, and increased levels of financial knowledge among participants. As these programs grow, they are faced with many of the same problems that have plagued financial education in a variety of populations: unclear metrics, lack of evidence-based practice, and a challenging population of learners.

Where Do Financial Literacy Programs Live on Campus?

As financial literacy programs on college campuses continue to evolve, there are many different models that have been developed. These models tend to be driven by where the program is housed administratively on a campus. Over time, these programs have been housed in a variety of areas of the university, which has created some issues related to how programs reach students and if they are able to effectively deliver education to their students. Although there remains a need for an administrative home for these programs, it is important to note that individual units need to partner to create a true campus-based approach to delivering financial education. And although programs have been, and currently are, housed in academic departments, financial aid, central service offices, and student affairs, it is important for areas not to see this as ownership, but rather as an organizational clearinghouse that works to build a community approach to assist students.

When financial education programs first began, they were most often found in academic programs. Traditionally, these programs were consumer sciences programs, or more specifically Certified Financial Planning (CFP) track programs, which were providing the academic education related to becoming a personal financial planner. These programs provided a great opportunity to reach out to the campus (and at times the broader community) with financial education resources and also provided additional reach for curricular offerings that may have already been provided by the department. At times, this extended (and still does extend) to include faculty members delivering financial workshops to students, but it increasingly became an opportunity for students to put into practice what they were learning in the classroom. This gave birth to the first wave of peer-to-peer financial coaching or counseling programs in some universities, including Texas Tech, University of Missouri, Kansas State University, and the Ohio State University. These programs allow for students to receive a great learning laboratory for their curricular studies by delivering group and one-on-one financial education, and these programs have great support as the faculty are very well versed in the topic areas. However, the largest challenge with these programs is that very often the academic side of the house can be siloed from the broader university community, specifically student affairs and enrollment management.

In recent years, these programs have begun to trend more in the direction of being housed within student support areas, and away from the academic core and academic programs (with some developing a hybrid model). This is partially attributable to the increased desire for institutions to provide financial education as a value-added service, coupled with a relatively small number of campuses that have CFP track educational curricula. These student support-housed programs were initially, and still are primarily, housed in financial aid offices on college campuses. This, on the surface, seems to make sense as the headlines about college student finances primarily focus on student loan debt and student loan default, both metrics that appear to fall most closely in line with the efforts that are normally undertaken by the financial aid office. A financial aid office provides a very positive home for these programs as they have access to significant information on the students' borrowing behaviors, have the ability to see trends related to student loan debt, and most importantly have the opportunity to change business practices in a way that helps to provide education and enhance students' financial wellness. There are, however, some challenges with programs being housed in these offices. Most notable of these is the inability to adequately provide holistic financial education without integrating into the broader campus community, including student affairs and academic affairs (for curricular offerings). This is often compounded by challenges in developing peer-to-peer programs (owing to the lack of interaction and connections to relevant students) and issues relating to finding a full-time (and qualified) employee to help manage the program.

The most recent trend—and the one we align with the closest at Ohio State—is to house a program within an office of student affairs. Although this may seem like an untraditional approach, there is an increased understanding of the role that student finances, and student financial stress, play in student success and potentially persistence. This understanding has led individuals to see a student's financial life as being interconnected with other areas of their lives, which in the case of Ohio State, manifests itself with financial being one of the nine dimensions of wellness that we work with. This view of finances as a holistic part of the student's experience lends itself nicely for financial literacy initiatives to be housed within a student affairs unit (including units focused on health and wellness). Additionally, student affairs units have built-in access to students—which is helpful when looking for avenues to reach students at different points in time. There remain challenges to the student affairs approach, including the lack of integration with the enrollment services units. This integration is integral to making sure that a financial literacy program is holistic in nature and that it helps to address some of the larger challenges facing students today, most notably student loan debt and management of cash flow throughout the semester.

Looking across the spectrum of administrative models, there is no perfect model. It is important to note this is a decision that varies by campus, and is often influenced by campus climate, campus politics, and other conversations about resource allocation. The one thing that remains consistent is that these programs rely on a campus-wide approach, and the development of a cross-unit team (sometimes referred to as an advisory board or task force) can go a long way toward developing partnerships and a true community-based approach to assisting students.

Rise in Financial Literacy Programs

Just as there are many different administrative homes for programs, there are also quite a few different ways that campuses deliver financial information to their student body. Research that has been conducted has recommended that an ideal financial education program provides resources online, in person through group workshops, and one-on-one coaching (Goetz et al., 2011). Seven different types of financial education are included in the following list, and many campuses use multiple approaches to best reach students (please note that this is not meant to be all inclusive, but rather to represent some of the most common types of programmatic offerings available today).

1. *Online Modules* are increasingly becoming a part of colleges and universities' answer to financial literacy challenges. Although their effectiveness has yet to be determined, they provide a great opportunity to reach students where they are (online) and assess their short-term learning in the process. These modules are being deployed as part of a prearrival checklist (before students come to campus) and during first-year seminar courses at many institutions across the country, which allows institutions to reach large numbers of students.
2. *Presentations/Workshops* have been in place for a long time in the financial education space, and they have been a long-standing favorite on college/university campuses. These sessions, lasting anywhere from 1 hour to 2 hours (or more), target a group of students through a one-to-many approach. Presentations are an effective way to boost reach numbers for a financial literacy program, but provide little evidence of long-term success for knowledge gain. Increasingly, there has been a trend toward providing just-in-time education through a workshop setting, which assists in making the content timely and relevant to the students who are on the receiving end of the information.

3. *One-on-One Appointments* provide a unique opportunity to engage a student in a very customized interaction related to their financial situation. This method of delivering education is gaining traction as it provides a safe space for the student to open up about issues that may be difficult to talk about and allows for customized education based on their current knowledge level and situation. Additionally, these appointments (on many campuses) are run through peer educators; often, these students are volunteer or low wage, which provides a scalable way to expand the service delivery. Furthermore, peers provide a comfortable setting in which students are able to expound on their situation as they relate to the individual who they are meeting with, trust them, and allow them to offer relevant solutions.
4. *Static Website Resources* are popular on financial aid websites across the country. These provide tips and links centered on a variety of financial topics. Although they may be updated, there is limited engagement with the student beyond the reading of the information on the page.
5. *Just-in-Time Education* is gaining traction in the field as a way to address the relevance of financial literacy information and deal with the potential for information overload. Higher education institutions have many opportunities to provide education at the point of sale (or decision) to help coach a student through the choices that they make and the effect that they have on their future. There is an opportunity for institutions to look further at their business practices as a place where they can integrate further just-in-time education including working with students receiving emergency support, those who may be at risk for drop for nonpayment, and those who may have a high percentage of unmet need.
6. *Curricular Offerings*, which are normally a semester long, provide an in-depth look at personal finance for interested students. Although these are very difficult to mandate students to take, there is an increasing reach of curricular offerings targeted at enhancing knowledge and providing a hands-on laboratory to put this knowledge to use. These courses can be expensive, as they are faculty intensive, but there are promising practices that may be emerging with a flipped classroom

format to drive increased online enrollment in such courses. It is also important to attempt to figure out ways for individual to receive credit for participation in such courses, including having them count towards general education requirements.

7. *Changes in Business Practices.* A recent trend, driven by Indiana University's recent changes to their financial aid practices, is to make adjustments to a college or university's business practices to help provide learning moments for students regarding their finances. This promising practice fits in with the public health environmental approach to building financial capability. Students may receive education as part of their financial aid selection process, a year-end financing report, or even as they navigate purchasing a dining plan. These changes, which are often low cost (to the institution), can go far in assisting the student body in developing positive financial habits.

One of the challenges presented by the numerous types of interventions, as well as the numerous homes for these interventions, is that there is no clearly agreed upon set of metrics for financial education programs. Oftentimes, much of the conversation on a campus—regardless of whether the campus is in the early or late stages of program development—revolves around determining what the metrics for a program may be. These may include reducing student borrowing, limiting student loan debt defaults, building financial knowledge, or supporting students in financial crisis.

Many of these metrics have multiple variables that contribute to the end outcome, making it difficult to assess the effectiveness of the interventions that are being provided. Additionally, it is necessary for campuses to be realistic about the life span of an intervention and how long a student can retain information from an intervention. These challenges are often presented by senior administrators early in the process of creating a program, and having a defined plan would be helpful in addressing these concerns.

It is important that, as we look to grow the field, we take a look at what metrics are achievable (in the confines of the intervention time frame) and how we can best articulate success, or failure, in a climate that is increasingly short-term outcomes based. Additionally, we need to take a look at ways to sustain these interventions to continually provide financial education beyond just a one-time intervention to our student body. Campuses have increasingly been looking at the complete life cycle of a student and identifying points in time to provide interventions and support students with nudges around their finances as well as identify students who may be at risk.

Need for Evidence-Based Interventions to Help Drive the Outcomes

One of the other biggest challenges facing the collegiate financial literacy field—and one that we will need to address as the prominence of these programs continues to grow in higher education—is the lack of evidence-based interventions that are proving to move the needle regarding our students' financial wellness. This problem is increasingly being seen across a variety of areas of the field, but is also increasingly being seen in higher education. There remain a few challenges that need to be worked out to ensure that we can move forward:

1. As previously mentioned, there is still no solid agreement in the field as to what the learning outcomes attached to interventions in the collegiate space should be. We know that it is unrealistic, as we push toward shorter times to degree, to require the completion of a full personal finance class for students, which means that the time of the intervention shifts significantly. Given the reduced face time with students, and the need to support students throughout their life cycle, we need to address best practices for providing education with targeted learning outcomes.
2. A secondary challenge is attributed to the relatively low levels of staffing associated with most financial literacy programs. There is little time to do assessment of programmatic interventions and develop evidence-based practice for the field. There is a need for the research community and the practitioner community to come together and take a long look at what we are doing and how we can assess the impact of our services. Ideally, additional grant funding would be made available to help support these endeavors.

Addressing these two challenges, which are by no means insurmountable, would go a long way in helping to work with university administration to increase resources related to financial literacy services. This increase in resources would help to provide students with increased access to services and more time for the enhancement of the interventions that are being offered.

What We're Doing at Ohio State University

The idea of financial education at Ohio State has been longstanding and continues to grow as we look forward to better serve our campus community. The program began in 2001 and was driven by a statewide effort to help promote the responsible use of credit products on college campuses. The initial financial literacy efforts included a set of presentations, as well as some one-on-one appointments, from a professional staff member, with the stated intent of developing healthy credit habits (note: this was before the banning of the marketing of credit products on our college campuses). In the early 2000s, this effort continued to expand, as the Office of Student Affairs (now the Office of Student Life) integrated financial wellness into the strategic plan for the Student Wellness Center, which in turn, integrated the financial wellness dimension into the official university wellness model. This point in time represents a transition in thought to view finances as a more holistic part of the student experience, and in line with that, interventions related to finances in a more holistic sense.

In 2006, the Office of Student Affairs entered into a partnership with the College of Education and Human Ecology to launch a peer-to-peer financial counseling program, called Scarlet and Gray Financial. This partnership allowed for the Student Wellness Center to utilize around six students, academically trained in financial planning, to deliver large group presentations and one-on-one financial interventions to students.

In 2012, we had the opportunity to further expand our services as we partnered with the Fisher College of Business to secure additional peer coaches to help deliver financial interventions to students. Currently, our team is made up of more than 40 peer coaches who volunteer to provide financial interventions to Ohio State students and deliver more than 150 presentations a year and more than 1400 one-on-one appointments to Ohio State students.

Our outreach program is structured to provide interventions that meet students where they are and address the students' needs at different points in their journey through Ohio State University. Our program, in line with research supporting the type of programmatic offerings that collegiate programs should offer, includes one-on-one in-person appointments, large group presentations, and a suite of online resources (Goetz et al., 2011). This allows us to reach students in a way that they prefer and effectively scale our reach to meet a wide variety of students. Additionally, two departments on campus offer curricular options for students who want a full-semester experience in personal finance. These programs attempt to raise awareness in our student body about their current financial situation, topics of financial relevance, and resources that are available to help them manage their finances.

Our one-on-one appointments are the primary tool that we use to help serve students and assist them in managing their financial lives. These appointments are 1-hour-long sessions conducted with a trained volunteer—a peer financial coach. The appointment process assists them in understanding their goals and their current financial standing, and developing an action plan to help manage their finances during their time at Ohio State and beyond. We are continuing to assess these efforts, but they show promise in increasing the students' awareness of their current financial situation and decreasing the stigma associated with talking about finances and their specific financial situation. As noted previously, we feel that peer-to-peer interaction has a very strong benefit in that students can relate to their peers, which lessens the stigma of seeking help and talking about their financial situation. For these appointments, we see students through a variety of avenues:

1. We see students who are mandated to attend as part of our Second Year Transformational Experience Program. This program is a multifaceted living learning and cocurricular experience that second-year students are able to opt into at our university. As part of this experience, students are required to complete a two-part financial wellness intervention (an online module followed by an in-person appointment). We see this as a very proactive opportunity to help expose students to

financial decision-making, including their current and future borrowing levels, and assist in making any behavior changes that they may want to make during their time at Ohio State.

2. Students who are interested in seeking help and/or referred from a variety of campus partners make up another segment of the student population that we see in our one-on-one appointments. These students may be proactive in learning more about their financial situation, or may be reactive in that they are in some form of financial trouble and they are now seeking assistance from our team. We rely heavily on the awareness gained through campus partners and presentations to help drive this form of traffic into our sessions.
3. Students who may be in some form of financial trouble can be referred as part of a mandated intervention integrated into our emergency loan program. A mandatory coaching appointment is included as part of the application process for an emergency loan. This reactive measure hopes to look at some of the factors that may be driving the need to apply for an emergency loan and prevent students from going into further financial trouble while developing habits to help them avoid the same pitfall in the future.

The second type of intervention comprises presentations and workshops that we are able to deliver across the campus. These targeted interventions, which normally last 1 hour, are presented to classes, student organizations, and other campus entities. These presentations are not intended to show significant gains in knowledge, attitudes, or behaviors, but rather are targeted at raising awareness about financial concepts, knowledge gaps, and resources available to students on Ohio State's campus. Additionally, the topics covered in these presentations are normally provided in a just-in-time fashion, meaning that the content is extremely relevant to certain experiences that the audience may be going through; these include the following sessions: Moving Off Campus, Repaying Student Loans, and Understanding Employee Benefit Packages.

The final type of intervention that we provide is a suite of online modules including two third-party software platforms, online calculators, and an iTunes U course. This allows students to access information in an on-demand fashion and helps us reach students where they are, online (at all hours of the day!). Every effort is made to try to combine an online intervention with an in-person follow-up appointment (such as the Second Year Transformation Experience Program), where possible.

Being housed in a Student Wellness Center, we have truly taken a public health approach to helping students develop financial wellness. We have done this by approaching the intervention environmentally, which lends itself to creating a campus-wide set of interventions, including empowering staff across the campus to have conversations about finances and helping to provide just-in-time financial education. Our Financial Wellness Advisory Board (which is made up of more than 32 campus partners from numerous offices) helps to keep this momentum moving and serves as the coordinating body for much of the work that is being done in so many departments across our campus. Our approach at Ohio State is truly a team approach and allows us to have a significant reach across the university.

National Landscape

The aforementioned growth in programs across the country has provided a great opportunity to help create a movement of collegiate financial wellness professionals across institutions of higher education. Unfortunately, owing to the decentralized nature of where these programs are administratively housed, the connection that has emerged across a variety of professional organizations lacks a focused concentration on the collegiate setting as it relates to financial literacy. We, at Ohio State, heard this broadly from colleagues across the country and engaged our counterparts at Indiana University in moving forward with a National Summit on Collegiate Financial Wellness to help bring together individuals in the field to share best practices and learn how to grow their efforts in their campuses.

The summit originated from the Statewide Summit on Collegiate Financial Wellness at Ohio State University in the spring of 2013. The 2013 Summit was modeled after an earlier statewide event that was coordinated to help facilitate the sharing of best practices in financial education in a collegiate setting within the state of Ohio. This event grew into the first annual National Summit on Collegiate

Financial Wellness, which was held during the summer of 2014 in Columbus, Ohio, at Ohio State University. More than 160 practitioners from more than 100 institutions and 34 states attended the first summit. The idea for the summit stemmed from a growing conversation in the field about the increase in the amount of programs aimed at enhancing the financial wellness of students, but there was a lack of opportunity to share best practices, connect with colleagues, and learn about emerging trends/issues. As financial wellness issues cross many different departments and units, the professional association for people doing this work is often divided, and we are hoping that the summit will provide a cross-discipline opportunity for professionals to connect, share best practices, and grow relationships that will lead to enhanced collaboration in the field.

Additionally, we are attempting to create a year-round forum for the positive exchange of ideas related to collegiate financial wellness. A listserv has been established to help provide this opportunity for dialogue across disciplines. All of this is an attempt to remove the silos that many individuals in the field feel as if they've been operating in. Because of the feedback we have received, the growth in programs across the country, and the need to align best practices, we are looking to continue to grow these resources for campuses in the future.

What Needs to Happen Next?

Given the relative infancy of the field of financial literacy in higher education, there is a tremendous room for growth and improvement as we look into the next few years.

1. Those currently in the field need to continue to come together, share best practices, and work toward an agreement on terminology and metrics. Many of the biggest challenges that we face related to gaining administrative buy-in and enhancing the programs that we currently have on college campuses revolve around these issues. It is important to note that these appear to be issues in the broader financial literacy community, and we should work to make sure that these conversations are not happening in a silo within higher education.

2. We need to continue testing interventions that are currently being put in place to see which ones have the potential to be successful. This process will require trial and error, and more importantly, support to allow for trial and error to take place. There is a need for us to be more intentional in how we engage and connect the academic community with practitioners to try to move these conversations forward and develop a research base around what works on a college campus.
3. College students, especially the current generation, are prime targets to be receiving enhanced just-in-time financial education messaging. Because of the increased level of information that they see, providing timely updates on financial information can help to inform their decision-making. Preferably, this would be extended to provide tools to allow individuals to receive this information at the time of purchase, leading to a more informed consumer.
4. It would be helpful to see an increased alignment with the K–12 financial education initiatives to create a continuum of financial education for students. There is a very strong, and well placed, effort to build financial education requirements in the K–12 space. This effort needs to continue (and be supported by the higher education community). Furthermore, it is imperative that these two areas work together to create a continuum of education, and also to prepare students to make a very large financial decision (often a decision entailing more than $100,000 in cost of attendance) in the best way that they possibly can.
5. We are continuing to seek a commitment to make the collegiate experience a time for financial development through a campus-wide effort to create an affordable experience. This requires the entire campus community to view financial wellness of their students as part of their job, and work tirelessly to help support it.

Although these challenges may seem numerous, there is a huge amount of excitement and energy in the collegiate financial wellness field at this time. There are many institutions that are making substantial commitments to help address a variety of concerns, and there is a great group of people leading these initiatives at various institutions across the country. As we provide an increasingly more accessible forum for these individuals to connect and share ideas, it is our hope that we can begin to address some of the largest challenges that have been impeding the growth in programs related to enhancing student financial literacy in the collegiate environment.

References

Goetz, J., Cude, B. J., Neilsen, R. B., Chatteriee, S., and Mimura, Y. (2011). College-based personal finance education: Student interest in three delivery methods. *Journal of Financial Counseling and Planning*, *22*(1), 27–42.

Trombitas, K. (2012). College students are put to the test: The attitudes, behaviors, and knowledge levels of financial education. Retrieved from Inceptia website: https://www.inceptia.org/about/resources/college-students-are-put-to-the-test/.

PART III

Post-College-Focused Financial Literacy Education

11

Financial Literacy in the Workplace

ERIC R. HECKMAN

Contents

The Human Resources (HR) representative told me, "please skip the basic course and give the advanced class since we have a lot of highly educated employees." I was teaching the advanced class, and when I got to the section on Roth Individual Retirement Accounts (IRAs), there were a lot of questions as to what is a Roth IRA or a Roth 401k. This was in 2010, a decade after the Roth IRA had come out and 2 years after this company added the Roth 401k option. Yes, it is true that the audience included people with PhDs in math and a lot of people with a Master's degree in computer science. How many classes in personal finance are required in the course curriculum for those degrees? None would be my guess. In my 23 years in the field of personal finance, I find no correlation that college education means the person has knowledge about asset allocation, investing, or financial products. In fact, the people who typically know more about these areas have gained that knowledge either from personal interest in the subject or they have had an actual class in the subject. I attended a very well respected private Jesuit college, Santa Clara University, and earned a BS in commerce with an emphasis in finance. How many required classes about investing or personal finance was I required to take? None. This may have changed since, but at the time I attended there was an elective class in personal finance—but nothing was required. We did economics that touched on personal financial decisions, but that was all. Most classes were about corporate finance. In fact, until I got into the financial services profession, the only educational exposure I truly received about anything financial was a few days in high school on checking accounts and budgets. Today's financial reality in our nearly pension-less society is that the employee is on his own for his financial future yet receives no training on financial matters.

When 401ks were taking hold in the 1980s and 1990s, most were a new thing for that company and for their employees. Most 401k providers would come out in person and hold talks about the plan and the investment options to educate the employees. In fact, this hands-on rollout of 401ks is what helped them get to be so popular. Fast forward to today, most large company 401ks are with a handful of firms, and in the Silicon Valley where I live, Fidelity is by far the largest. Fidelity, like most large financial firms, focus their time on providing website content, 800 numbers, and maybe offer a physical location here and there. It is in fact rare to see these firms providing much hands-on, classroom-style education at the employers' locations. Even if they do offer the class, it is focused solely on the 401k and no other financial topic.

Once, employee benefits were just the one health insurance plan your company had and your pension. Today, especially in high-tech firms, there are multiple health insurance options, health or flexible spending accounts, dependent care, educational reimbursement, legal plan, employee stock purchase plan (ESPP), restricted stock units (RSUs), stock options, 401k, and Roth 401k if not more. Just about every one of these options has unique tax ramifications and all affect your financial plan. HR departments focus on giving employees information on what each item is, when they need to make a decision, and how to enroll. The item missing from all this is strategy. This leaves the employees asking questions such as "When is it good to fund a FSA versus an HSA? How do the RSUs or ESPPs affect your taxes? When should you sell your company's stock? How do all these benefits fit in with my spouse's options and our other investments we have outside of work place benefits?"

The U.S. Department of Labor (DOL) Code Section 404c discusses the liability and roles of the employer and plan's fiduciary for retirement plans. The subject of employee education is not truly defined, but there is some clarity found in the DOL's Employee Benefits Security Administration (EBSA) publication, *Meeting your Fiduciary Responsibilities*; in the Tips section, it asks:

> If participants make their own investment decisions, have you provided the plan and investment related information participants need to make informed decisions about the management of their individual accounts? Have you provided sufficient information for them to exercise control in making investment decisions?*

The brochure also discusses the fact that you can hire someone to educate the employees or have someone provide general information about financial matters. If you as the employer choose to not hire someone, then the employer must educate their employees in financial matters and that duty is usually delegated to Human Resources (HR) or the training department. Actually, a lot of companies solely depend on their 401k provider to offer this education which typically is just a website and maybe some videos on it. If the HR department takes on the challenge, the HR person has to determine how to offer the education. They may need to teach the class which begs the question of how much financial education do they have to lead it? In fact, they run the risk of offering financial advice or tax advice without any license or legal authority to provide it. This could be a liability for the firm and the HR person is acting in a fiduciary capacity for the company. To avoid this, when the training is run by HR people, the talk is typically more of 'here are

* http://www.dol.gov/ebsa/pdf/meetingyourfiduciaryresponsibilities.pdf. Meeting Your Fiduciary Responsibilities Brochure, Feb 2012, U.S. Department of Labor EBSA.

the details of the plans, benefits and features' and nothing more. This type of discussion doesn't answer the question the employee has which is 'how this is relevant to my situation and why is this something I should act on?'

How can a company get help doing this and avoid the HR person running the education? A lot of commissioned stockbrokers, insurance agents, and other financial advisors would be happy to stand in front of a bunch of confused employees and sell them their products. The problem with this is the fact they are motivated for the employees to fund items they sell, which may lead to the recommendation of funding less to company benefits. This advice may not be in the best interest of the employee, and it doesn't have to be in their best interest legally speaking. The vast majority of so-called financial advisors are under the suitability rule. This means that their advice must be appropriate in terms of age, timeline, and goals, but it does not need to be in the employee's best interest. Anyone who works as a registered representative of a broker–dealer falls under this rule. There is a movement to try to change this standard of advice, but it is not law as of this time. If someone is a registered investment adviser, then they must use the fiduciary rule and recommend what is in the best interest of the client. If you are in HR and want to bring someone in to provide financial educational content to your employees, I would make sure they are working for a registered investment adviser. They are usually called investment advisory representatives. The final outsourcing option is to bring in a firm that specializes in doing financial classes for employees. There are a few firms that do this and charge a fee to the employer. There are also some nonprofit firms that do it for no cost like the not-for-profit organization called Financial Knowledge Institute (FKI) that I founded.

I was frustrated by the lack of financial knowledge that most of my clients had and wanted to do something about it. That is when I founded FKI. I like to teach people about financial matters, and I wanted to give back to my community. I designed the courses to go over all aspects of a person's financial situation so we have more than just a registered investment advisor talk but a realtor, mortgage broker, attorney, and near tax time a certified public accountant (CPA). I think this knowledge is so critical, yet it is rarely represented in a nonsales type of format. With social media ads, e-mail, TV, radio, and Internet financial ads being shown to us constantly, we get immune to it and stop paying attention to the actual good information we need to have. Every so often there may be an investment class offered through an adult education community center; however, these are often taught by commissioned brokers, not registered investment advisers. One of the biggest barriers to learning about financial matters is quite simply ourselves.

Most of us tend to ignore things we must do and focus on things that are fun to do. Should you estimate how much you need to retire in 20 years or should you spend time determining the best stops to make on your summer vacation? Sadly, an Edward Jones study shows that of the time people spend studying financial decisions, they spend 28% on vacation compared to 25% on retirement. Under 35, that percentage for retirement drops to 9%.* Ironically, people prefer to plan their vacation before their retirement even though a better retirement could mean more vacations in the end. So the dilemma is how to motivate people to take the time to prioritize their financial futures. This is why I feel educating employees at the worksite is so helpful.

We all get too many e-mails every day. There is no real way to force financial education on people, although having it conveniently located at the workplace and knowing their employer is okay with them taking the time to learn makes the subject and the process of learning more appealing to them. I think the big value of utilizing someone who is qualified to cover an entire subject is the employee gets a much more in-depth understanding of the issues and how it relates to them.

One of the most difficult items to tackle is the fact that the vast majority of companies are made up of employees from their 20s to their 70s now. There are several stages of financial eras that people go through over their lifetimes. In your 20s and 30s, it is more about starting to save, buying a home, making a lifestyle, and starting a family. In your 40s and 50s, it may be more focused building wealth,

* Americans Spend More Time Planning Vacations Than Retirement, by Libby Kane, June 26, 2014, available at http://www.businessinsider.com/americans-plan-vacations-over-retirement-2014-6.

improving your career, and helping a child through college. In your 60s and older, it is certainly focused on retirement. So the challenge is how to speak to all of these stages of life in one class. Anyone older than 30 understands the basics of a 401k and maybe even stock options and a lot of the other benefits. In their 40s and 50s as income increase, the employees often need to look beyond the 401k to determine other ways to save and invest while lowering the tax implications. In their 60s, they are trying to navigate the maze of issues such as when to start social security, how to generate income instead of just return, and how to minimize the taxes as they start to withdraw these funds. I think the best way to handle this varying degree of issues depends on the size of the firm.

Big companies, 500 employees and up, can host financial sessions on site and start with a basic class and then host classes more targeted to issues at certain ages such as those about to retire age 55 and older. Those nearing retirement are in the need of the most financial help. Most firms under this size need to offer just a more basic course, and then those employees need to ask the instructor for more help for their situation. Some smaller firms entice employees to attend by offering a free lunch of pizza or burritos, which does indeed increase attendance. In my experience, these free meals bring more employees to the workshop; however, it does not result in more people asking to get more information from the instructor!

Ideally, anyone teaching these classes who are qualified to lead them should offer some sort of follow-up time, either in person or on the phone. Like sex and politics, money is not looked upon as a good subject to discuss in the workplace. Although I find there are a lot of questions at these sessions, there are many more that people are afraid or uncomfortable to ask in a public setting especially with people they see every day. I think the instructor should be expected to offer some time, at least 30 minutes or so as a free consultation. Most people who are qualified to teach these classes do have a business of their own. In my case, I love to talk about finance and teach people how to connect different financial concepts and products together. I do need to make money, too, so my consultations are free, but often there are a few people who do want to hire me to help them more. I think the prime focus of people giving these talks should be to give back to the community while raising the profile of their firm in their community second. Personally, I meet with anyone who asks for a consultation usually in person or at least on the phone. The issues they want to discuss are all over the place as you would expect in a varied age range company. I have had simple items such as "should I pay off my college loan or fund my 401k?" At the same company, I will hear "what is the best allocation for the $1 million I have invested?" The instructor should be ready and, I believe, willing to at least talk to all these people. If you are in HR looking for a firm to do this education, verify they will do this for your employees.

Does It Actually Lead to Better Employees or Happier Employees?

Personally, I haven't done any research or analysis to track this beyond the response sheets that we ask attendees to complete. The vast majority state they do enjoy the classes and did learn new information that they can use. In terms of meeting with the employees, often the knowledge gap is huge. One firm that I hold several classes in each year is a typical high-tech Silicon Valley firm in that they need a lot of highly qualified engineers. A large number of these engineers come from varied countries around the world. Contributing to the U.S. Social Security system while having benefits and maybe assets in another country certainly complicates financial planning. Often, they have no knowledge of U.S. retirement plans. Even without these international issues, there are a lot of other time specific questions people have that need to be answered. The financial crisis of 2008–2009 was one of these times when people had a lot more concerns than usual. With the stock markets crashing, educating employees to the pros and cons of trying to time the market or react too quickly in hindsight proved to be critical information. We had to add extra classes for the Real Estate team after the real estate bubble popped to go over short sales and loan modifications. I really don't think companies, especially during critical financial times, realize the stress that their workforce is under and how educating these people about the options out there can really improve their mindset and their productivity. In a Society for Human Resource Management (SHRM) study quoted in Purchasing Power's "Money Smarts: Helping employees make the grade" white paper, HR professionals state that 50% of employees are affected by financial stress, and 47% let it affect their productivity.*

In one of my first talks at Synopsys, I had a gentleman come and ask if we could meet. Let's call him Paul for this story. Paul had lived in Wisconsin many years earlier, and his old financial advisor stopped keeping in touch with him. A lot of his investments were not allocated for his timeline and goals, plus he had no income plan for retirement. We were able to combine a lot of accounts to make his plan simpler and easier for him to track, lower his risk, and acquire a guaranteed lifetime income annuity from an insurance company because he has no pension. We also had the attorney who was doing the estate planning talks set him up with a trust for him and his wife. Soon after, his wife moved to New York to care for her dying mother. After her mother passed away, she stayed out there because she was starting to have medical issues of her own. We were able to set up an income for her there that made it easier for Paul to keep working. Now, several years later, his wife needs more continuous care and after a year of paying for round the clock care, Paul did retire and will be moving there soon.

Paul is a very dedicated employee whose main hobby is work. I am convinced that had Synopsys not brought FKI in to do these talks, he would have been forced to leave the firm several years sooner. He would have had some legal issues around caring for his wife and he would not have been able to retire without worry. His story is not unusual by any means and illustrates why I think financial education at the workplace is so critical. Paul would not have gone out and tried to find what he needed or educate himself on money matters because it did not interest him. Having the financial educational resources come to him and making them easy to use empowered him to take financial control of his life while staying a productive employee.

---

* https://www.purchasingpower.com/sites/all/files/attachments/employer-resource/Money-Smarts-Financial-Wellness-Education-White-Paper-Feb2015.pdf. Purchasing Power's White Paper Money Smarts: Helping Employees Make the Grade, Feb 2015.

A Guide to How to Offer This at Your Company

The desire to offer this at your company can come from the person in charge of the 401k, the Human Resources department wanting someone else to do the training, or the employee wanting the education to come to them. This guide will be directed at each one of them and provide a timeline for the program and some best practices.

First, it makes sense to give a basic overview of the process. Typically, the person doing the talk will provide either bullet points for an e-mail invitation or maybe a PDF document discussing the class. Then, it is the duty of someone at the hosting company to promote it to the employees. That may fall on the HR department or a training person to promote it in whichever manner they usually communicate benefit changes. The hosting company will provide a conference room for the workshop. The speaker will then host the talk, usually 45 minutes with 15 minutes for questions and answers. I find after the talks there are four to five people who want to ask a question in private, too. There should be response sheets to get employee feedback on the program and for the employee to request a consultation. These forms should be kept by the presenter, but a scan of all of them should be sent to the host company contact so they can judge the value of the program. For the employees who ask for a consultation, the speaker will typically meet them at their company and the employee is in charge of finding a conference room or a location to meet. After the talk, the host company contact does not have any more to do. Beyond the free consultation, then it is up to the employee to engage the financial planner, attorney, or real estate professional at their own cost.

So where do you go to find someone to do these talks? Talking to other companies in your area that have offered a program like this is a good place to start. Online searches for companies that offer employee education are another avenue to investigate. FKI is my nonprofit, but there are several others. You may even get calls from financial firms trying to offer this, but be very wary in such cases. As I mentioned earlier this may be product and sales driven and not educational based. I would highly recommend attending another talk the person gives at another firm just to make sure you like it. What you really don't want is a long sales pitch or education disguised just to lead to the sale of a specific product. Speaking for my own nonprofit, we do professional background checks on every professional—financial, legal, real estate, or tax—to make sure they are truly an expert in their field and there are not any legal issues. These background checks are fairly easy to perform; the following list includes a few options:

Financial Advisers with securities license—http://www.FINRA.org
Registered Investment Advisers—http://www.sec.gov/investor/brokers.htm
Insurance agents are state by state—http://www.naic.org/state_web_map.htm
Attorneys are state by state—http://www.americanbar.org
Mortgage brokers—http://www.nmlsconsumeraccess.org/
Realtors® are also state by state—http://www.realtors.com

The side benefit of this background check is you are verifying that they are licensed to provide advice in their area. However, the financial profession is a very difficult field to easily tell who can do what and give advice in which areas. As I mentioned earlier, I feel the person should either be a registered investment advisor or a person who works for one—which means "Investment Advisory Representative" will be on their business card. They also could be a Certified Financial Planner® (CFP®) or a Chartered Financial Consultant® (ChFC®).

Building a Team of Professionals

Often the motivation to present the talks may be led by the financial advisor, and he or she may have others but may not. I think doing some education is better than nothing so that is okay, but I feel it is really best to offer all the programs. The attorney may be someone whom the advisor knows or if the company offers a legal plan benefit then look to an attorney from there. The realtor should be a full-time realtor and ideally one who is a good presenter. The mortgage person I feel really should be a mortgage broker, not a mortgage person from a bank. The reason for this is only a mortgage broker can offer all the products on the market and truly discuss them all. If someone is from Wells Fargo say, they can only really discuss their products—making the talk less educational. So why this team approach? It can be really sad if someone gets their finances and investments in order but loses them to a lawsuit or legal fees because they were not set up correctly. Mortgages are a huge part of a person's financial plan so taking care of this area is also important. There is also something to be said for building momentum and keeping it going once someone tries to tackle all these areas. This series of classes does that.

Timeline and Process

The timeline and process can be

- HR finds an organization to provide the financial education.
- HR verifies the organization is credible and the speaker has the right credentials.
- HR and the organization decide on a date or series of dates to hold the talks.
- The organization provides HR with an e-mail and a PDF flyer for the event.
- HR sends out the information and does a follow-up RSVP days before the talk.
- HR assists the organization in having the conference room ready.
- The speaker presents the workshop and allows for questions.
- The speaker has the employees complete a response sheet for feedback and to offer consultations.
- The speaker sends a copy of those response sheets to HR for them to review.
- The speaker conducts free consultations for those who asked for it.
- The employee, after getting the consultation, then has the option of paying for the professional's service or not.
- HR and the organization ideally host a series of talks and will have the next date set.
- If there are more than 500 employees, the organization should come in at a minimum of every 6 months, but ideally this should occur quarterly.

Tips for the Employer

As mentioned earlier, Labor Code 404c requires the company to provide financial information about their plan. If you are the plan trustee, the more information you can offer to your employees the lower the liability can be for you.

Tips for Human Resource or Training Personnel

As I mentioned earlier, having your 401k provider come in annually or more often is a must. You are their client and you pay them for this service so they owe it to you to educate your employees. Then you need to determine if you want to go over the other benefits. The risky part here is where to cross the line on providing financial or tax advice when discussing ESPP or RSUs or options. Even some of the tax-advantaged items such as a health savings account (HSA) could expose your company to some risk by providing and educating the employees about the tax savings. This is why I really recommend finding an objective—ideally nonprofit—firm to give the educational talks. On the plus side, it is one less item you need to do. Promoting the meeting is critical to it being effective, and that can only be done by HR or training. I would check out the organization, review the speakers, and at least do some Google searches of them. Going to one of their talks at another company is also a good practice. SHRM or your local HR society may have contacts or connections to not only find an organization to provide the talks but also someone who has hosted them to find out what the company thought of the experience. Often and with good reason HR professionals are wary of bringing in outside firms to talk to their employees. Remember, offering these talks should ultimately do several positive things for your company: meet the education requirements for the 401k, take the workload off HR, increase the use of company benefits, and create happier, less stressed employees.

The HR professional should fully explain to the speaker all the benefits the firm offers. You would be surprised that even if you had sent out a lot of information on a particular benefit, many employees have still not read about it. Also, you want to make sure the speaker talks about the benefits you offer, not ones that your firm doesn't provide. The Roth 401k is a great and often misunderstood benefit, yet if the speaker talks a lot about it and you don't offer it, it could cause issues. So before the workshop, make sure the speaker knows your benefits.

Tips for All Speakers

Your role in providing these talks must always be first and foremost to educate the audience. Providing relevant, current information on the overall financial situation as well as information on how the company's benefits can be used to help the employee's financial picture. Remember that for most talks the audience will be of varying ages so the initial talks should be more basic and general in nature. One key item I feel is the need to educate the employee about items not available at work and how they relate to workplace benefits and steps they need to set those up on their own. The most popular class we offer is one on Alternative Investments, especially as few people are exposed to some of these. Doing talks geared at the "Retirement Red Zone," as it is often called—those 10–15 years from retiring—is also a good idea so you can focus more on Social Security claiming strategies, and focus on income and other risks that retirees face.

Best Practices for a Professional to Develop a Speaking Team

As the talks really need to start with the basic financial talk, often this process is spearheaded by the financial advisor. To really serve the company in the best way possible, I feel it is critical to offer workshops that fully cover an employee's financial issues. This must include legal, tax, real estate, and debt. This means you need to find professionals who not only want to present these workshops but that he or she would also offer some free consultations. The other important item is that they are good presenters. The legal and tax fields are not typically known for their charismatic personalities. First and foremost, they must really know the subject; but just as important, they need to be able to explain it in an easy-to-understand, nonjargon manner. For the real estate and mortgage fields, there are often people in those industries that do it part time. I feel it is more appropriate to have a speaker who is working full time in the industry, actively know the status of that industry, and can offer the postworkshop consultations. Another reality is that the speaker's commitment level may drop off over time or, depending on the workload, may no longer work for that person. Be prepared for the possibility of needing to replace the speakers every year or two.

Best Practices for Financial Professional

There are several things the workshop should not include. It goes without saying that anything about sex, politics, and religion should never be discussed. However, there have been times when Congress or the president have threatened to change something or not renewed something that expired, which has financial ramifications. Discussing these issues is important, but adding a political viewpoint should be avoided. I once had an attorney start to discuss the president and included a joke about the president. That led to two negative comments on the response sheet that got to the HR person and had us remove him from the speakers list. Talks should not refer to a specific product such as XYZ annuity, but discussing annuities and how they fit in is fine. People love to give predictions on the stock markets and employees will often ask the speaker to provide them but these, too, should be avoided. Reviewing economic facts such as the Federal Reserve has stated they want to raise interest rates or that the United States has an $18 trillion deficit is relevant and appropriate. Avoid using the situation you mention to recommend the audience to "go buy this fund or stock" because of it. Any presentation you give should be reviewed by your compliance department or compliance officer. It is also wise for someone with a securities license to list this volunteer activity as an OBA or Outside Business Activity. I am a registered investment adviser, and this activity is listed on my Securities and Exchange Commission Form ADV Part 2 to make sure it is fully disclosed to the investing general public. Many financial advisers have a minimum net worth or asset level before they will work with someone and that cannot be used in a situation like this. Remember you are volunteering your time so make sure to help anyone who asks for it. If they require more help but they are not a fit for your firm, then please refer them to someone who can help them.

Best Practices for the Attorney

More and more companies are offering a voluntary legal benefit plan through a firm such as Hyatt. It is really beneficial if the attorney doing the presentations is a member of the company's benefit provider, as it makes it easier for the employee to follow through to implement the documents they need because it is covered by the plan. Often, the first class should be just focused on wills, living trusts, and other basic documents while explaining who needs which. Then a more advanced class can be offered to go more into other types of issues such as estate planning. I find a shockingly low number of people have really done what is needed on the legal side of a financial plan. These talks can be very useful to inspire action on the employee's part but need to be done in an easy-to-understand, nontechnical manner.

Best Practices for the Realtor–Mortgage Broker Team

The reason I feel these workshops should be done as a team is because questions always come up for the other professional. "Should I buy my first home" question always requires a discussion on mortgages rates, down payments, and tax and other issues. The Great Recession really led to more talks on debt management and credit scores, which the mortgage speaker should be well versed on. Even today, there are still short sales happening where you need the expertise of the mortgage broker along with the realtor. Real estate and interest rates are, of course, very location-specific. Because that is the case, you do need to cover current market conditions and offer some sort of assessment to the audience. These consultations are not always as needed because people may have a good mortgage with a home they don't plan on selling. Offering credit score checks or home valuations are a way to give some real value to a consultation.

Best Practices for a Tax Speaker

This speaker could be a CPA or an enrolled agent. Ideally, this speaker needs to possess some skills to turn what many see as a boring subject into something interesting. During tax time, these talks should not be done. It is way too busy for the speaker and even the audience. I find that toward year-end is a great time to hold these talks. People have a good idea of what happened income-wise and still have time to act before the year is over. The year-end is when someone should always spend time tax planning because once the year is over, so too are most tax saving options. The talk should focus on ways the company's benefits can help the employee lower their taxes now or in the future and any other moves to make before the year's end. Any new tax laws should be covered as well as any ones that are expiring. This class really can be done just once a year and is not totally required for a successful program, but I highly recommend it.

Best Practices for the Employee

Okay, providing guidance to the employee may seem odd, but I think there is a lot of value that employees overlook and options they miss. Obviously, just by attending these programs, you become more knowledgeable about personal finance issues that you may or may not have right now. Even if some don't apply to you, they may apply to your parents or friends and you would then be in a position to help them. Asking questions during the workshop is helpful, too. If any part of the talk applies to your situation, then please take the speaker up on their free consultation. The brief time you spend with this professional is usually valued at $200 an hour or so.

The Future for Financial Education in the Workplace

I think the financial turmoil of the 2000s have led some to have a desire to learn more and take control of their finances. But for many, the financial setbacks they encountered have made them avoid financial items they don't understand. Tax laws are always changing at the whims of Congress, and often these have a direct impact on investments. Employee benefits change and employees themselves change companies. All of these changes need to be accounted for in the employee's financial plan but may very well be overlooked caused by a busy life. For some, they will turn to the increasing number of online financial tools such as "robo advisers" websites that provide asset allocation advice, but that is it. It is very difficult to have a computer look at a tax return to find investment mistakes, to analyze a trust, or for an online mortgage quote to coordinate with the person's financial plan. These tasks will still need the use of a human, and that human needs a way to connect to employees. I think on-site workplace education can really solve many of these issues and even some of the issues that plague the United States in general. Had someone discussed their new home purchase and 100% mortgage with their financial adviser, they may have not bought a house that was later foreclosed upon. Although I don't think companies should be required to provide this education, they will find that their employees' lower stress levels can lead to higher productivity. Startups would still be smart to offer these because they may not be able to afford great benefits, and a program like this can help employees determine other ways to save on taxes. I did a talk at the software firm Workday while it was still a private company. The employees who attended were able to save thousands on taxes by learning more about the stock options they owned and how to decrease taxes on them. One employee hired me later after they went public and we were able to provide some advice on an investment that allowed him to lower his tax bill by over tens of thousands of dollars on his IPO (initial public offering) stock sales.

There are numerous benefits to these programs for all parties involved and only one real drawback—letting a salesman in. This is the main objection or fear of the firm talking to the education provider. It is truly a valid fear but one that can be easily overcome. Getting references from other firms where they have spoken, watching one of the presentations, and of course making sure they are qualified to speak on the matter should avoid any issues. The question I don't think enough firms ask is "What happens to my star employees" in times of financial crisis. The Great Recession of 2008–2009 is one thing, but a financial crisis can be as simple as an employee losing their spouse suddenly and there not being enough life insurance to take care of their kids. That employee may need to take a job closer to home and leave the firm. It doesn't even need to be that dramatic. An employee could be struggling paying off debt while not knowing there is a loan option on the 401k. The stress that employees feel every day could easily translate into lower productivity, and it could be even worse if they are in a financial role, which may tempt them to embezzlement or theft. I know a small firm that had this exact situation, and their chief financial officer is now in jail and the firm's losses from the theft led to it eventually shutting its doors.

Workplace financial education is a simple yet powerful way that corporate America can make real progress in increasing financial literacy, increasing employee stability, and even achieving productivity gains.

United States Government Accountability Office

Testimony
Before the Subcommittee on Financial Institutions and Consumer Credit, House Committee on Financial Services

or Release on Delivery
xpected at 10:00 a.m. ET
/ednesday, April 30, 2014

FINANCIAL LITERACY

Overview of Federal Activities, Programs, and Challenges

Statement of Alicia Puente Cackley, Director
Financial Markets and Community Investment

O-14-556T

April 30, 2014

FINANCIAL LITERACY

Overview of Federal Activities, Progress, and Challenges

GAO Highlights

Highlights of GAO-14-556T, a testimony before the Subcommittee on Financial Institutions and Consumer Credit, House Committee on Financial Services

Why GAO Did This Study

Giving Americans the information they need to make effective financial decisions is key to their financial well-being. The federal government plays a role in promoting financial literacy, which encompasses financial education—the process by which individuals improve their knowledge and understanding of financial products, services, and concepts. The federal role evolved with the creation of the Commission in 2003 and CFPB in 2010.

This testimony provides an (1) overview of federal financial literacy activities and agency roles, and (2) update on the progress made in addressing GAO's recommendations in this area. This testimony is largely based on and partially updates a July 2012 report (GAO-12-588) and relevant portions of GAO's annual duplication reports (GAO-11-318SP and GAO-12-342SP). For those reports, GAO reviewed and analyzed relevant reports, plans, and websites related to federal financial literacy efforts, and interviewed staff of 17 federal agencies and of nonprofit organizations. To update selected information, GAO spoke with staff and reviewed materials from CFPB and the Department of the Treasury.

What GAO Recommends

GAO makes no new recommendations in this testimony, but reiterates its July 2012 recommendations for the Commission to identify options for consolidating federal financial literacy efforts and address the allocation of federal resources in its national strategy. The Commission agreed with these recommendations.

View GAO-14-556T. For more information, contact Alicia Puente Cackley at (202) 512-8678 or cackleya@gao.gov.

What GAO Found

In its July 2012 report, GAO identified 16 significant financial literacy programs c activities among 14 federal agencies in fiscal year 2010, and 4 housing counseling programs among 2 agencies and a federally chartered entity. As of April 2014, 3 of the financial literacy programs and 1 housing counseling prograr no longer existed or no longer received funding, and no new federal programs had been added, according to staff representing the Financial Literacy and Education Commission (Commission). The creation of the Bureau of Consumer Financial Protection (known as CFPB) in 2010 added a significant new player, with offices devoted to financial education broadly and to educating servicemembers, older Americans, and students specifically. The multiagency Commission, created in 2003, coordinates among federal agencies and betwee federal agencies and state, local, nonprofit, and private entities. Commission responsibilities also include developing national strategies for improving financia literacy and proposing means of eliminating overlap and duplication among federal activities. Finally, since 2008 three Presidential advisory councils relatec to financial literacy have had charges that include fostering partnerships among private, nonprofit, and government entities.

GAO has observed improvements or successes in four areas—coordination, partnerships, delineating CFPB's role, and evaluation tools—but significant wor remains to be done in one major area—determining the most effective and efficient allocation of federal resources.

- **Coordination**. Coordination among federal agencies has improved in recen years, largely due to the role of the Commission. Recent efforts include a research clearinghouse and a coordinated initiative on youth education.
- **Partnerships.** The Commission has continued to build and promote partnerships. Several initiatives have partnered with academics, nonprofits, and other entities.
- **Delineating CFPB's role.** To help avoid unnecessary overlap, CFPB has further delineated its role in financial literacy efforts, discussing respective roles with federal agencies that have overlapping responsibilities and signin agreements on cooperation and areas of focus.
- **Evaluation tools.** The Commission and CFPB have helped develop and disseminate evaluation tools to assess outcomes and effectiveness of financial literacy programs. CFPB also contracted with a company to develc metrics and outcome measures and with a nonprofit to evaluate and report financial education programs and activities.

However, further progress is needed to help ensure effective allocation of fede financial literacy resources and avoid unneeded overlap. In 2012, GAO concluded the Commission was best placed to consider consolidating federal efforts, which could help ensure the most efficient and effective use of federal resources. The Commission's national strategy could serve as a mechanism t identify those resources and how they might be allocated, but has not yet done so. Without recommendations on resource allocations, policymakers lack information to avoid overlap and help ensure the most efficient and effective u of federal funds.

Chairman Capito, Ranking Member Meeks, and Members of the Subcommittee:

I am pleased to be here today to discuss financial literacy and related federal programs as part of Financial Literacy Month 2014. Giving Americans the information they need to make effective financial decisions is key to their financial well-being. Moreover, financial markets function best when consumers understand how financial service providers and products work and how to choose among them. The federal government has played a key role in addressing financial literacy, and this role has evolved in recent years, particularly with the creation of the Financial Literacy and Education Commission (Commission), which coordinates federal efforts, in 2003 and the Bureau of Consumer Financial Protection (known as CFPB) in 2010.[1]

My testimony today is largely based on and partially updates a report we issued in July 2012 that examined federal financial literacy activities, including identifying agencies, programs, activities, and coordination.[2] It also incorporates work we conducted in recent years as part of our overall efforts to address fragmentation, overlap, or duplication in the federal government.[3] This testimony provides (1) a brief overview of federal financial literacy programs, activities, and agency roles; and (2) a progress update on recommendations we made in this area.

[1]The commission currently comprises 22 federal entities; its Chair is the Secretary of the Treasury and its Vice Chair, as established in the Dodd-Frank Wall Street Reform and Consumer Protection Act (Dodd-Frank Act), is the Director of CFPB. Pub. L. No. 111-203, § 1013(d)(5), 124 Stat. 1376, 1971 (2010) (codified at 20 U.S.C. § 9702(c)(1)(C).

[2]See GAO, *Financial Literacy: Overlap of Programs Suggests There May Be Opportunities for Consolidation*, GAO-12-588 (Washington, D.C.: July 23, 2012). This report addresses (1) what is known about the cost of federal financial literacy activities; (2) the extent and consequences of overlap and fragmentation among federal financial literacy activities; (3) what the federal government is doing to coordinate its financial literacy activities; and (4) what is known about the effectiveness of federal financial literacy activities.

[3]See GAO, *2012 Annual Report: Opportunities to Reduce Duplication, Overlap and Fragmentation, Achieve Savings, and Enhance Revenue*, GAO-12-342SP (Washington, D.C.: Feb. 28, 2012); and *Opportunities to Reduce Potential Duplication in Government Programs, Save Tax Dollars, and Enhance Revenue*, GAO-11-318SP (Washington, D.C.: Mar. 1, 2011). See also *GAO's Action Tracker*, a publicly accessible online website of the areas and actions presented in these reports, which provides progress updates and assessments of the actions we suggested for Congress and executive branch agencies.

To address the objectives of our 2012 report, we reviewed and analyzed relevant reports, surveys, agency strategic plans, performance and accountability reports, websites, budget justifications, performance data, and evaluations related to federal financial literacy efforts. We interviewed staff of 17 federal agencies and staff of nonprofit organizations. In addition, we collected cost information from congressional appropriations, agency budget justifications, and staff estimates. To update selected information for this testimony, we spoke with staff and reviewed related materials of CFPB and the Department of the Treasury (Treasury).

We conducted the work on which this testimony is based in accordance with generally accepted government auditing standards. Those standards require that we plan and perform the audit to obtain sufficient, appropriate evidence to provide a reasonable basis for our findings and conclusions based on our audit objectives. We believe that the evidence obtained provides a reasonable basis for our findings and conclusions based on our audit objectives. More details on our scope and methodology are provided in each of the related products.

Background

Financial literacy, sometimes referred to as financial capability, has been defined as the ability to use knowledge and skills to manage financial resources effectively for a lifetime of financial well-being. Financial literacy encompasses financial education—the process by which individuals improve their knowledge and understanding of financial products, services, and concepts. However, to make sound financial decisions, individuals need to be equipped not only with a basic level of financial knowledge, but also with the skills to apply that knowledge to financial decision making and behaviors.

Efforts to improve financial literacy can take many forms. These can include one-on-one counseling; curricula taught in a classroom setting; workshops or information sessions; print materials, such as brochures and pamphlets; and mass media campaigns that can include advertisements in magazines and newspapers or on television, radio, or billboards. Many entities use the Internet to provide financial education, which can include information and training materials, practical tools such as budget worksheets and loan and retirement calculators, and interactive financial games.

)verview of Federal ʼrograms or Activities •n Financial Literacy nd Agency Roles

In our 2012 report, we identified 16 significant financial literacy programs or activities administered by the federal government in fiscal year 2010.[4] As shown in table 1, there were 16 programs or activities among 14 federal agencies and 4 housing counseling programs (which can include elements of financial education) among 2 federal agencies and a federally chartered nonprofit corporation. The programs and activities covered a wide range of topics and target audiences and used a variety of delivery mechanisms. While we have not done a comprehensive update of this federal program summary since 2012, representatives of Treasury and CFPB—which represent the Chair and Vice Chair of the Commission, respectively—told us that as of April 2014, 3 of the 16 financial literacy programs and 1 of the 4 housing counseling programs either no longer existed or no longer received funding, and no new federal programs had been added since fiscal year 2010.[5]

[4]We defined "significant" as those financial literacy and education programs or activities that were relatively comprehensive in scope or scale and included financial literacy as a key objective rather than a tangential goal. See page 12 of GAO-12-588 for additional information about our criteria for significant programs or activities.

[5]The discontinued programs are the Social Security Administration's Financial Literacy Research Consortium, the Department of Education's Excellence in Economic Education Program and Financial Education for College Access and Success Program, and Treasury's Financial Education and Counseling Pilot Program.

Table 1: Description and Target Audience for Significant Federal Financial Literacy and Housing Counseling Programs and Activities, Fiscal Year 2010

Financial Literacy			
Agency	**Program or activity**	**Description**	**Target audience**
Board of Governors of the Federal Reserve System	Division of Consumer and Community Affairs and Office of Public Affairs	Up-to-date web resources on regulatory changes regarding financial products and services, calculators, and information and tools on terms and disclosures for credit card accounts, overdraft protection programs, gift cards, and credit scores. Website offers resources for teachers and students of various ages and knowledge levels through educational games, classroom lesson plans, online publications, and multimedia tools.	Adult consumers and students
Bureau of Consumer Financial Protection	Office of Financial Education and other offices	The Offices of Financial Education, Servicemember Affairs, Fair Lending and Equal Opportunity, and Financial Protection for Older Americans develop and implement initiatives to educate and empower consumers in general and specific target groups to make informed financial decisions.	Consumers, servicemembers and their families, and individuals who are 62 or older
Department of Agriculture	Family and Consumer Economics programs	The National Institute of Food and Agriculture provides funding to land-grant colleges and universities and to state and county extension offices to support research and education, including outreach events related to personal financial topics.	Youth, rural families, elderly, and other financially vulnerable populations
Department of Defense	Personal Financial Management Program (located within Family Support Centers)	Personal financial managers on military installations provide financial education programs and counseling services designed to help servicemembers reach their financial goals. Services range from consultation on financial management, budgeting, and saving, to debt reduction strategies, consumer advocacy and complaint resolution, financial workshops, retirement planning, housing issues and referrals, and education programs for youth and teens.	Servicemembers and their families
Department of Education	Excellence in Economic Education Program	Awarded competitive grant to an organization that conducted activities, and made subgrants to other organizations, to improve the quality of student understanding of personal finance and economics.	Students in kindergarten through grade 12
	Financial Education for College Access and Success Program	Supported state-led efforts to develop, implement, and evaluate personal finance instructional materials and teacher training intended to aid students in making financial aid decisions in relation to postsecondary education.	Students in middle ar high-school—generally, grades 6-1
Department of Health and Human Services	National Education and Resource Center on Women and Retirement Planning	Provides women access to a one-stop gateway on retirement, care giving, health, and planning for long-term care.	Low-income women, women of color, and women with limited English proficiency
Department of Labor	Saving Matters Retirement Savings Education Campaign	Workplace campaign to promote retirement savings and understanding of federal retirement law using interactive web tools, print publications, website, public service announcements, seminars, workshops, videos, and webcasts.	Employees, employers, and sma businesses
	Wi$eUp	Eight-module financial education curriculum targeting women. Topics include money basics, credit, saving and investing, insurance, retirement planning, and financial security. Offered online or in a classroom setting.	Generation X and Y women

Agency	Program or activity	Description	Target audience
Financial Literacy			
Department of the Treasury	Office of Financial Education and Financial Access (now incorporated into the Office of Consumer Policy)	A variety of financial literacy activities and staff support for the Financial Literacy and Education Commission and MyMoney.gov (website on federal financial literacy resources).	All populations
Federal Deposit Insurance Corporation	Money Smart Financial Education Program	Eleven-module financial education curriculum for adults designed to enhance basic financial skills and create positive banking relationships, available in nine languages. Eight-module version is available for young adults. The curriculum is available in instructor-led, computer-based instruction, and podcast (Mp3) formats.	Low- to moderate-income adults outside the financial mainstream and youth ages 12-20
Federal Trade Commission	Division of Consumer and Business Education	Multimedia resources covering topics such as credit, credit repair, debt collection, job hunting, job scams, managing mortgage payments, avoiding foreclosure rescue scams, and identity theft.	Consumers
Office of the Comptroller of the Currency	Consumer education activities	Websites, consumer advisories, public service announcements, community outreach, and print and radio advertisements aimed at educating consumers about banking and other financial issues.	Consumers
Office of Personnel Management	Retirement Readiness NOW	Retirement education strategy designed to provide information that will help federal employees plan for retirement and calculate the investment needed to meet retirement goals.	Federal employees
Securities and Exchange Commission	Office of Investor Education and Advocacy	Provides information to help individual investors evaluate current and potential investments, make informed decisions, and avoid fraud.	Investors
Social Security Administration	Financial Literacy Research Consortium	Supported 2-year cooperative agreements with Boston College, RAND Corporation, and the University of Wisconsin to develop innovative materials and programs to help Americans plan for a secure retirement.	Adults preparing for retirement
Housing Counseling and Foreclosure Mitigation			
Department of Housing and Urban Development	Housing Counseling Assistance Program	Certifies and oversees housing counseling providers. Provides competitive grants to approved housing counseling agencies that provide pre- and post-purchase counseling, assistance to renters, homeless populations, and those seeking to resolve mortgage delinquency. Counseling may take place in person, over the telephone, or with a self-study computer module or workbook.	Low- to moderate-income families
Department of the Treasury	Financial Education and Counseling Pilot Program	Awarded competitive grants to organizations to provide financial education and counseling to prospective homebuyers.	Prospective homebuyers
NeighborWorks America	National Foreclosure Mitigation Counseling Program	Provides competitive grants to housing counseling agencies to provide one-on-one counseling services for foreclosure prevention.	Homeowners at risk of foreclosure
	Other housing counseling activities	Provides expendable grants for which housing counseling is an eligible activity.	Current and prospective homeowners

Source: GAO analysis based on information from federal agencies and interviews with agency staff.

Note: According to Treasury and CFPB representatives, as of April 2014, four of the programs listed in the table either no longer existed or no longer received funding—the Social Security Administration's Financial Literacy Research Consortium, the Department of Education's Excellence in Economic Education Program and Financial Education for College Access and Success Program, and Treasury's Financial Education and Counseling Pilot Program.

CFPB, which became operational in 2011, has a primary role in addressing financial literacy.[6] For example, its Office of Financial Education provides opportunities for consumers to access financial counseling; information on understanding credit products, histories, and scores; information on savings and borrowing tools; and help in developing long-term savings strategies and wealth building. The Office of Servicemember Affairs helps educate servicemembers and their families to make better-informed decisions on consumer financial products and services, monitors complaints, and coordinates efforts among federal and state agencies regarding consumer protection measures. The Office of Financial Protection for Older Americans develops goals for programs that provide financial literacy and counseling to help seniors recognize the warning signs of unfair, deceptive, or abusive practices. CFPB's Office of Fair Lending and Equal Opportunity plays a role in providing education on fair lending. Finally, CFPB created an Office for Students to address complaints and questions regarding student loans.

In addition to specific programs and activities directed at consumers, federal agencies participate in coordination or umbrella efforts. For instance, the 22-member Commission coordinates among other federal agencies and between federal agencies and state, local, nonprofit, and private entities. Also among Commission charges are developing a national strategy for improving financial literacy and proposing means of eliminating overlap and duplication among federal financial literacy activities. Furthermore, since 2008, three Presidential advisory councils related to financial literacy have been tasked, in part, with creating partnerships among federal, state and local, nonprofit, and private entities. Most recently, the President's Advisory Council on Financial

[6]CFPB was created by the Dodd-Frank Act, which specified the creation of the bureau's Office of Financial Education and its role in promoting financial literacy. Pub. L. No. 111-203, § 1013(d),124 Stat. 1376, 1970 (2010) (codified at 12 U.S.C. § 5493(d)).

Capability for Young Americans was created by Executive Order in June 2013.[7]

[F]ederal Agencies [A]ddressed Several [G]AO [R]ecommendations, [B]ut More Progress [N]eeded on [D]etermining the Best [A]llocation of [R]esources

Our reviews of federal financial literacy efforts in recent years have examined, among other things, issues of fragmentation and overlap; the effectiveness of programs; coordination across multiple agencies and between federal, state, local, and nonprofit entities; the potential consolidation of programs or activities; and the need to improve the Commission's national strategy for financial literacy.[8] The following summarizes progress in four areas in which we have observed improvements or successes—coordination, partnerships, CFPB's role, and evaluation tools—and one major area in which we believe significant work remains to be done—determining the most effective and efficient allocation of federal resources.

[C]oordination

Overall, coordination on financial literacy among federal agencies has improved in recent years. As noted earlier, multiple federal agencies have significant financial literacy initiatives. Because of the crosscutting nature of financial literacy, it would be difficult, if not impossible, for one agency alone to address the issue, but coordination among agencies is clearly essential. In our 2006 report, we noted that the Commission enhanced communication and collaboration among agencies involved by creating a single focal point for federal agencies to come together on the issue of financial literacy. In 2011, we suggested that the Commission enhance its

[7]Exec. Order No. 13646, 78 Fed. Reg. 39159 (June 28, 2013).

[8]For examples of GAO work on financial literacy, see *Financial Literacy: Enhancing the Effectiveness of the Federal Government's Role*, GAO-12-636T (Washington, D.C.: Apr. 26, 2012); *Highlights of a Forum: Financial Literacy: Strengthening Partnerships in Challenging Times*, GAO-12-299SP (Washington D.C.: Feb. 9, 2012); *Financial Literacy: A Federal Certification Process for Providers Would Pose Challenges*, GAO-11-614 (Washington D.C.: June 28, 2011); *Financial Literacy: The Federal Government's Role in Empowering Americans to Make Sound Financial Choices*, GAO-11-504T (Washington D.C.: Apr. 12, 2011); *Financial Literacy and Education Commission: Progress Made in Fostering Partnerships, but National Strategy Remains Largely Descriptive Rather Than Strategic*, GAO-09-638T (Washington, D.C: Apr. 29, 2009); *Financial Literacy and Education Commission: Further Progress Needed to Ensure an Effective National Strategy*, GAO-07-100 (Washington, D.C.: Dec. 4, 2006); and *Highlights of a GAO Forum: The Federal Government's Role in Improving Financial Literacy*, GAO-05-93SP (Washington D.C.: Nov. 15, 2004).

efforts to coordinate federal activities, such as by exploring further opportunities to strengthen its role as a central clearinghouse for federal financial literacy resources.[9] The Commission has addressed this suggestion with a number of additional coordinating activities taken since 2012. These have included an internal web portal to allow federal agencies involved in financial literacy efforts to share information and resources; publication of *2012 Research Priorities and Research Questions* to help coordinate federal research efforts; a clearinghouse of federal research and data on financial literacy; an initiative for coordinating among federal agencies' activities and resources to help parents and teachers prepare children and young adults for financial success; and support to the Office of Personnel Management in promoting agency strategic plans for employee retirement readiness.[10]

Partnerships

The Commission has continued to build on its progress in promoting partnerships among federal and nonfederal sectors. Given the wide arra of state, local, nonprofit, and private organizations providing financial literacy programs, it is essential to leverage private-sector resources an coordinate federal activities with resources at the community level. We suggested in 2011 that the Commission build on progress it had made i promoting partnerships among the federal, state, local, nonprofit, and private sectors.[11] The Commission has taken actions that address this suggestion. Examples of recent Commission work with nonfederal sect include the above-mentioned 2012 research report, which the Commission developed in partnership with academic researchers, nonprofit financial educators, and other nonfederal stakeholders, and a initiative to create a community of practice among educators and policymakers. The commission also continued to work in collaboration with the President's Advisory Council on Financial Capability (which ended in 2013) and the new President's Advisory Council on Financial Capability for Young Americans. The new council primarily consists of members from the private and nonprofit sectors, as did the former council.

[9]GAO-11-318SP, pp. 151-154.

[10]Financial Literacy and Education Commission, *2012 Research Priorities and Resear Questions: Financial Literacy and Education Commission Research and Evaluation Working Group* (Washington, D.C.: May 2012).

[11]GAO-11-318SP, pp. 151-154.

Delineation of CFPB's Role

CFPB has taken steps to delineate and distinguish its role in financial literacy from that of other federal agencies. In 2011 and 2012, we noted that some of CFPB's financial literacy responsibilities appeared to overlap with those of other federal agencies, and we recommended that steps be taken to ensure a clear delineation of these agencies' respective roles and responsibilities. In its response to a draft of the 2012 report, CFPB neither agreed nor disagreed with this recommendation.[12] We believe this recommendation has been implemented. Since 2012, financial literacy staff from CFPB and Treasury have continued to hold regular meetings to discuss respective roles, and with the emergence of CFPB, Treasury reorganized the structure of its own financial education efforts, incorporating them into the broader Office of Consumer Policy. Similarly, as of March 2013, CFPB's Office of Servicemember Affairs was meeting monthly with staff responsible for financial literacy at the Department of Defense, and the agencies developed two Joint Statements of Principles to help delineate their roles and responsibilities. Furthermore, CFPB's Office of Financial Protection for Older Americans finalized a memorandum of understanding with the Federal Trade Commission in January 2012 to help cooperation on consumer education efforts and promote consistent messages. CFPB's Office of Students developed a memorandum of understanding with the Department of Education designed, in part, to clarify respective areas of focus in providing education on student loans and financial aid.

Evaluation Tools

We suggested in 2011 that the Commission and CFPB take steps to develop and disseminate a standard set of evaluation tools or benchmarks to help assess the outcomes and effectiveness of financial literacy programs.[13] This suggestion has been addressed. In September 2012, CFPB signed a contract with the Corporation for Enterprise Development to develop a set of metrics and outcome measures for assessing the success of financial literacy programs. That work is ongoing. CFPB also contracted with The Urban Institute to evaluate financial education programs and increase understanding of which interventions can improve financial decision-making skills in consumers. An interim report was published in January 2014 and the final report is

[12]GAO-12-588.

[13]GAO-11-318SP, pp. 151-154.

expected around April 2015, according to CFPB representatives.[14] In addition, the Commission's 2012 study on research priorities sought ways of making the best use of limited dollars to address the most important questions facing the field of financial literacy.

Allocation of Federal Resources

We continue to believe that further progress is needed in one area—helping to ensure that the most effective and efficient allocation of federal financial literacy resources occurs and avoids unneeded overlap. As noted earlier, federal financial literacy and housing counseling resources are spread across many agencies, the result of both legislation and programs evolving to address a variety of populations or topics. While our prior work uncovered no duplication, some agencies or programs do have overlapping goals and activities, raising the risk of inefficiency and underscoring the importance of coordination. The creation of CFPB added a new player to the mix. While progress has been made in coordinating with federal agencies that have overlapping financial literacy responsibilities, we noted in 2012 that the creation of CFPB signaled an opportunity for reconsidering more broadly how federal financial literacy efforts are organized. In particular, some consolidation of these efforts could help ensure the most efficient and effective use of federal resources. As a result, in July 2012 we recommended to Treasury and CFPB that the Commission identify for federal agencies and Congress options for consolidating federal financial literacy efforts into the agencies and activities that are best suited or most effective.[15] Commenting on a draft of that report, Treasury agreed with the recommendation, and subsequently CFPB noted its agreement as well, but the recommendation has not been implemented.

The Commission's national strategy could serve as one mechanism for determining how federal resources might best be allocated among programs and agencies, but it does not yet do so. The Commission issued its first national strategy in 2006. We reported that it was a useful first step, but largely descriptive rather than strategic, and only partially included certain key characteristics such as a description of resources

[14]Brett Theodos, Margaret Simms, Claudia Sharygin, et al., *Rigorous Evaluation of Financial Capability Strategies: Why, When, and How* (Washington, D.C.: January 2014); prepared by The Urban Institute for the Bureau of Consumer Financial Protection.

[15]GAO-12-588.

required to implement the strategy.[16] In December 2010, the Commission issued its second and most recent national strategy.[17] While it discussed the consumer education resources the federal government makes available to consumers, it still did not address the level and type of resources needed to implement the strategy, or review the budgetary resources available for financial literacy efforts and how they might best be allocated. In July 2012, we recommended that the Commission revise its national strategy to incorporate clear recommendations on the allocation of federal financial literacy resources across programs and agencies.[18] Treasury and CFPB agreed with this recommendation as well, but have not yet implemented it.

We acknowledge that the governance structure of the Commission presents challenges in addressing resource issues: it relies on the consensus of multiple agencies, has no independent budget, and no legal authority to compel members to act.[19] However, the Commission can identify resource needs, make recommendations, and provide guidance on how Congress or federal agencies might allocate scarce federal resources for maximum benefit. Without a clear description of resource needs, policymakers lack information to help direct the strategy's implementation. And without recommendations on resource allocations, policymakers lack information to avoid overlap and help ensure the most efficient and effective use of federal funds.

Chairman Capito, Ranking Member Meeks, and Members of the Subcommittee, this concludes my prepared statement. I would be happy to answer any questions that you may have at this time.

ontact and Staff cknowledgments

If you or your staff have any questions about this testimony, please contact me at (202) 512-8678 or cackleya@gao.gov. Contact points for our Offices of Congressional Relations and Public Affairs may be found

[16] Financial Literacy and Education Commission, *Taking Ownership of the Future: The National Strategy for Financial Literacy* (Washington, D.C.: April 2006) and GAO-07-100.

[17] Financial Literacy and Education Commission, *Promoting Financial Success in the United States: National Strategy for Financial Literacy 2011* (Washington, D.C.: December 2010).

[18] GAO-12-588.

[19] GAO-09-638T.

on the last page of this statement. GAO staff who made key contributions to this testimony include Jason Bromberg (Assistant Director); Bethany Benitez; Juliann Gorse; Barbara Roesmann; and Rhonda Rose.

Index

Note: Page numbers ending in "f" refer to figures. Page numbers ending in "t" refer to tables.

U

V

W

X

For Product Safety Concerns and Information please contact our EU
representative GPSR@taylorandfrancis.com Taylor & Francis Verlag GmbH,
Kaufingerstraße 24, 80331 München, Germany

Printed and bound by CPI Group (UK) Ltd, Croydon, CR0 4YY
08/05/2025
01864453-0001